HIGHER

CHEMISTRY
2006-2010

First exam published in 2006.

Published by Bright Red Publishing Ltd, 6 Stafford Street, Edinburgh EH3 7AU

tel: 0131 220 5804 fax: 0131 220 6710 info@brightredpublishing.co.uk www.brightredpublishing.co.uk

ISBN 978-1-84948-134-2

A CIP Catalogue record for this book is available from the British Library.

Bright Red Publishing is grateful to the copyright holders, as credited on the final page of the book, for permission to use their material.

Every effort has been made to trace the copyright holders and to obtain their permission for the use of copyright material.

Bright Red Publishing will be happy to receive information allowing us to rectify any error or omission in future editions.

[BLANK PAGE]

FOR OFFICIAL USE

Total
Section B

X012/301

NATIONAL
QUALIFICATIONS
2006

TUESDAY, 30 MAY
9.00 AM – 11.30 AM

CHEMISTRY
HIGHER

Fill in these boxes and read what is printed below.

Full name of centre

Town

Forename(s)

Surname

Date of birth

Day   Month   Year   Scottish candidate number   Number of seat

Reference may be made to the Chemistry Higher and Advanced Higher Data Booklet (1999 edition).

SECTION A—Questions 1–40 (40 marks)

Instructions for completion of **Section A** are given on page two.

For this section of the examination you must use an **HB pencil**.

SECTION B (60 marks)

1   All questions should be attempted.

2   The questions may be answered in any order but all answers are to be written in the spaces provided in this answer book, **and must be written clearly and legibly in ink**.

3   Rough work, if any should be necessary, should be written in this book and then scored through when the fair copy has been written. If further space is required, a supplementary sheet for rough work may be obtained from the invigilator.

4   Additional space for answers will be found at the end of the book. If further space is required, supplementary sheets may be obtained from the invigilator and should be inserted inside the **front** cover of this book.

5   The size of the space provided for an answer should not be taken as an indication of how much to write. It is not necessary to use all the space.

6   Before leaving the examination room you must give this book to the invigilator. If you do not, you may lose all the marks for this paper.

SCOTTISH
QUALIFICATIONS
AUTHORITY

SECTION A

Read carefully

1 Check that the answer sheet provided is for **Chemistry Higher (Section A)**.

2 For this section of the examination you must use an **HB pencil** and, where necessary, an eraser.

3 Check that the answer sheet you have been given has **your name**, **date of birth**, **SCN** (Scottish Candidate Number) and **Centre Name** printed on it.

  Do not change any of these details.

4 If any of this information is wrong, tell the Invigilator immediately.

5 If this information is correct, **print** your name and seat number in the boxes provided.

6 The answer to each question is **either** A, B, C or D. Decide what your answer is, then, using your pencil, put a horizontal line in the space provided (see sample question below).

7 There is **only one correct** answer to each question.

8 Any rough working should be done on the question paper or the rough working sheet, **not** on your answer sheet.

9 At the end of the exam, put the **answer sheet for Section A inside the front cover of your answer book**.

Sample Question

To show that the ink in a ball-pen consists of a mixture of dyes, the method of separation would be

        A chromatography

        B fractional distillation

        C fractional crystallisation

        D filtration.

The correct answer is **A**—chromatography. The answer **A** has been clearly marked in **pencil** with a horizontal line (see below).

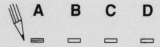

Changing an answer

If you decide to change your answer, carefully erase your first answer and using your pencil, fill in the answer you want. The answer below has been changed to **D**.

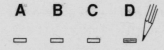

1. A negatively charged particle with electronic arrangement 2, 8 could be

   A a fluoride ion

   B a sodium atom

   C an aluminium ion

   D a neon atom.

2. Which of the following exists as diatomic molecules?

   A Helium

   B Methane

   C Carbon monoxide

   D Sodium chloride

3. When an atom **X** of an element in Group 1 reacts to become X^+

   A the mass number of **X** increases

   B the charge of the nucleus increases

   C the atomic number of **X** decreases

   D the number of filled energy levels decreases.

4. A mixture of magnesium chloride and magnesium sulphate is known to contain 0·6 mol of chloride ions and 0·2 mol of sulphate ions.

   What is the number of moles of magnesium ions present?

   A 0·4

   B 0·5

   C 0·8

   D 1·0

5. When a gas was bubbled through dilute hydrochloric acid, the pH increased.

   The gas could have been

   A ammonia

   B hydrogen

   C methane

   D sulphur dioxide.

6. The graph below shows the change in the concentration of a reactant with time for a given chemical reaction.

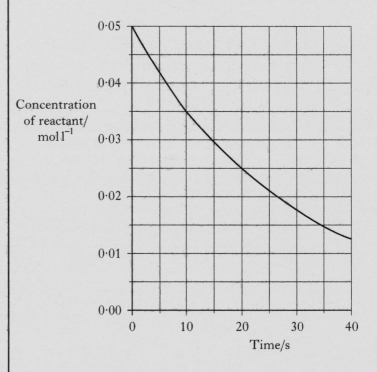

   What is the average rate of this reaction, in $mol\,l^{-1}\,s^{-1}$, between 10 and 20 s?

   A $1\cdot0 \times 10^{-2}$

   B $1\cdot0 \times 10^{-3}$

   C $1\cdot5 \times 10^{-2}$

   D $1\cdot5 \times 10^{-3}$

7. A small increase in temperature results in a large increase in rate of reaction.

   The **main** reason for this is that

   A more collisions are taking place

   B the enthalpy change is lowered

   C the activation energy is lowered

   D many more particles have energy greater than the activation energy.

[Turn over

8. When copper carbonate reacts with excess acid, carbon dioxide is produced. The curves shown were obtained under two different conditions.

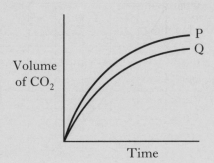

The change from **P** to **Q** could be brought about by

A   increasing the concentration of the acid

B   decreasing the mass of copper carbonate

C   decreasing the particle size of the copper carbonate

D   adding a catalyst.

9. $63 \cdot 5$ g of copper is added to 1 litre of 1 mol l^{-1} silver(I) nitrate solution.

Which of the following statements is always true for this reaction?

A   The resulting solution is colourless.

B   All the copper dissolves.

C $63 \cdot 5$ g of silver is formed.

D   One mole of silver is formed.

10. A potential energy diagram is shown.

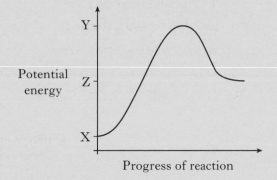

What is the activation energy (E_A) for the forward reaction?

A   Y

B   Z – X

C   Y – X

D   Y – Z

11. A group of students added 6 g of ammonium chloride crystals to 200 cm^3 of water at a temperature of 25 °C.

The enthalpy of solution of ammonium chloride is $+13 \cdot 6 \text{ kJ mol}^{-1}$.

After dissolving the crystals, the temperature of the solution would most likely be

A   23 °C

B   25 °C

C   27 °C

D   30 °C.

12. The spike graph shows the variation in successive ionisation energies of an element, **Z**.

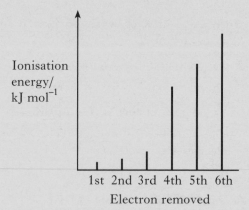

In which group of the Periodic Table is element **Z**?

A   1

B   3

C   4

D   6

13. Which equation represents the second ionisation energy of magnesium?

A $Mg^+(g) \rightarrow Mg^{2+}(g) + e^-$

B $Mg(g) \rightarrow Mg^{2+}(g) + 2e^-$

C $Mg(s) \rightarrow Mg^{2+}(g) + 2e^-$

D $Mg^+(s) \rightarrow Mg^{2+}(s) + e^-$

14. Carbon dioxide is a gas at room temperature while silicon dioxide is a solid because

   A van der Waals' forces are much weaker than covalent bonds

   B carbon dioxide contains double covalent bonds and silicon dioxide contains single covalent bonds

   C carbon-oxygen bonds are less polar than silicon-oxygen bonds

   D the relative formula mass of carbon dioxide is less than that of silicon dioxide.

15. Which of the following contains the same number of atoms as 16 g of helium?

   A 16 g of methane

   B 16 g of oxygen

   C 17 g of ammonia

   D 20 g of argon

16. A one carat diamond used in a ring contained 1×10^{22} carbon atoms.

   What is the approximate mass of the diamond?

   A 0·1 g

   B 0·2 g

   C 1·0 g

   D 1·2 g

17. $C_2H_6(g) + 3\frac{1}{2}O_2(g) \rightarrow 2CO_2(g) + 3H_2O(\ell)$

   When 20 cm^3 of ethane was sparked with 100 cm^3 of oxygen, what was the final volume of gases?

   All volumes were measured at atmospheric pressure and room temperature.

   A 40 cm^3

   B 70 cm^3

   C 100 cm^3

   D 130 cm^3

18. During the manufacture of petrol, the process of reforming is used to produce a petrol with a higher percentage of molecules which are

   A aromatic

   B larger

   C unbranched

   D unsaturated.

19. An ester has the following structural formula:

   $CH_3CH_2CH_2COOCH_2CH_3$

   The name of this ester is

   A propyl propanoate

   B ethyl butanoate

   C butyl ethanoate

   D ethyl propanoate.

20. Which of the following compounds has isomeric forms?

   A C_2H_3Cl

   B C_2H_5Cl

   C C_2HCl_3

   D $C_2H_4Cl_2$

21. Which of the following structural formulae represents a primary alcohol?

   A
$$CH_3 - CH_2 - CH_2 - \overset{\displaystyle H}{\underset{\displaystyle OH}{C}} - CH_3$$

   B
$$CH_3 - CH_2 - \overset{\displaystyle H}{\underset{\displaystyle OH}{C}} - CH_2 - CH_3$$

   C
$$CH_3 - \overset{\displaystyle CH_3}{\underset{\displaystyle OH}{C}} - CH_2 - CH_3$$

   D
$$CH_3 - \overset{\displaystyle CH_3}{\underset{\displaystyle CH_3}{C}} - CH_2 - OH$$

[Turn over

22. What product(s) would be expected on dehydrating the following alcohol?

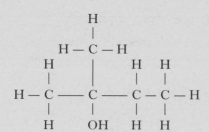

A 2-methylbut-2-ene only

B 2-methylbut-2-ene and 2-methylbut-1-ene

C 2-methylbut-1-ene only

D 3-methylbut-1-ene and 2-methylbut-1-ene

23. Which of the following statements about benzene is correct?

A Benzene is an isomer of cyclohexane.

B Benzene reacts with bromine solution as if it is unsaturated.

C The ratio of carbon to hydrogen atoms in benzene is the same as in ethyne.

D Benzene undergoes addition reactions more readily then hexene.

24. Which of the following consumer products is **least** likely to contain esters?

A Flavourings

B Perfumes

C Solvents

D Toothpastes

25. Which of the following types of bond is broken during hydrolysis of a polyamide?

A C — N

B C = O

C N — H

D C — C

26. Synthesis gas can be made by

A reacting ethene gas with steam

B burning carbon in excess air

C burning methane gas in excess air

D reacting methane gas with steam.

27. Some recently developed polymers have unusual properties.

Which polymer is soluble in water?

A Poly(ethyne)

B Poly(ethenol)

C Biopol

D Kevlar

28. The flow diagram shows two steps in the preparation of a feedstock for use in making a plastic.

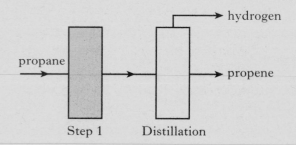

The reaction taking place during step 1 is an example of

A cracking

B oxidation

C dehydration

D hydrogenation.

29. What type of reaction is involved in the conversion of vegetable oils into "hardened" fats?

A Condensation

B Hydration

C Hydrogenation

D Polymerisation

30. Which of the following could be a monomer for the formation of a protein?

A $HOOC - CH_2 - COOH$

B $HOCH_2 - CHOH - CH_2OH$

$$\underset{\underset{\displaystyle C \quad CH_3 - CH - COOH}{|}}{NH_2}$$

$$\underset{\underset{\displaystyle D \quad CH_3 - CH - COOH}{|}}{OH}$$

31. Consider the reaction pathways shown below.

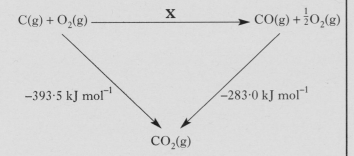

According to Hess's Law, the enthalpy change for reaction **X** is

A $+110 \cdot 5 \ kJ \ mol^{-1}$

B $-110 \cdot 5 \ kJ \ mol^{-1}$

C $-676 \cdot 5 \ kJ \ mol^{-1}$

D $+676 \cdot 5 \ kJ \ mol^{-1}$.

32. In a reversible reaction, equilibrium is reached when

A   molecules of reactants cease to change into molecules of products

B   the concentrations of reactants and products are equal

C   the concentrations of reactants and products are constant

D   the activation energy of the forward reaction is equal to that of the reverse reaction.

33. Gaseous iodine and hydrogen were reacted together in a sealed container.

$$I_2(g) + H_2(g) \rightleftharpoons 2HI(g)$$

Which of the following graphs shows the pressure inside the vessel as the reaction proceeds at constant temperature?

A

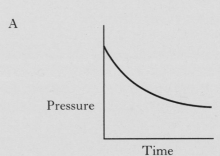

B

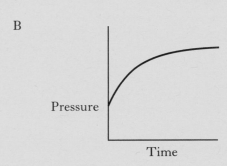

C

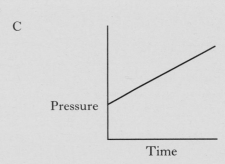

D

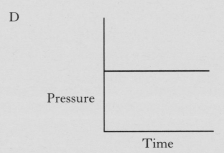

[Turn over

34. Which of the following is the same for equal volumes of $0.1 \, \text{mol} \, l^{-1}$ solutions of sodium hydroxide and ammonia?

A The pH of solution

B The mass of solute present

C The conductivity of solution

D The number of moles of hydrochloric acid needed for neutralisation

35. Which of the following would cause the pH of an acid solution to be changed from 2 to 4?

A Diluting $100 \, \text{cm}^3$ of solution to $200 \, \text{cm}^3$ with water

B Evaporating $100 \, \text{cm}^3$ of solution until $50 \, \text{cm}^3$ remain

C Diluting $1 \, \text{cm}^3$ of solution to $100 \, \text{cm}^3$ with water

D Evaporating $100 \, \text{cm}^3$ of solution until $1 \, \text{cm}^3$ remains

36. Which of the following compounds dissolves in water to give an alkaline solution?

A Sodium nitrate

B Potassium ethanoate

C Ammonium chloride

D Lithium sulphate

37. $HgCl_2(aq) + SnCl_2(aq) \rightarrow Hg \, (\ell) + SnCl_4(aq)$

What ion is oxidised in the above redox reaction?

A $Sn^{2+}(aq)$

B $Sn^{4+}(aq)$

C $Hg^{2+}(aq)$

D $Cl^-(aq)$

38. The unlabelled line on the graph below was obtained by plotting the mass of copper metal deposited against charge passed during the electrolysis of a melt of copper(I) chloride.

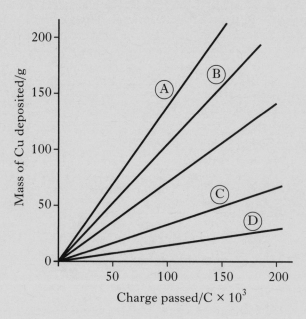

If a melt of copper(II) chloride was electrolysed, which line would correspond to the mass of copper metal deposited against charge passed?

39. ^{14}C has a half life of 5600 years. An analysis of charcoal from a wood fire shows that its ^{14}C content is 25% that of living wood.

How many years have passed since the wood for the fire was cut?

A 1400

B 4200

C 11 200

D 16 800

40. A radioisotope of thorium forms protactinium-231 by beta-emission.

What is the mass number of the radioisotope of thorium?

A 230

B 231

C 232

D 235

Candidates are reminded that the answer sheet MUST be returned INSIDE the front cover of this answer book.

Marks

SECTION B

All answers must be written clearly and legibly in ink.

1. The melting and boiling points and electrical conductivities of four substances are given in the table.

Substance	Melting point/°C	Boiling point/°C	Solid conducts electricity?	Melt conducts electricity?
A	92	190	no	no
B	1050	2500	yes	yes
C	773	1407	no	yes
D	1883	2503	no	no

Complete the table below by adding the appropriate letter for each type of bonding and structure.

Substance	Bonding and structure at room temperature
	covalent molecular
	covalent network
	ionic
	metallic

(2)

[Turn over

Marks

2. The elements in the second row of the Periodic Table are shown below.

Li	Be	B	C	N	O	F	Ne

(a) Why does the atomic size decrease crossing the period from lithium to neon?

1

(b) Diamond and graphite are forms of carbon that exist as network solids.
Name a form of carbon that exists as discrete molecules.

1

(c) Use the electronegativity values to explain why nitrogen chloride contains pure covalent bonds.

1

(3)

Marks

3. Catalytic converters in car exhaust systems convert poisonous gases into less harmful gases.

   (a) Two less harmful gases are formed when nitrogen monoxide reacts with carbon monoxide.

   Name the **two** gases produced.

   1

   (b) The catalyst is made up of the metals platinum, palladium and rhodium.

   Explain what happens to molecules in the exhaust gas during their catalytic conversion to less harmful gases.

   You may wish to draw labelled diagrams.

   2

   (3)

   [Turn over

Marks

4. The enthalpy of combustion of methanol (CH_3OH) can be determined from measurements using the apparatus shown.

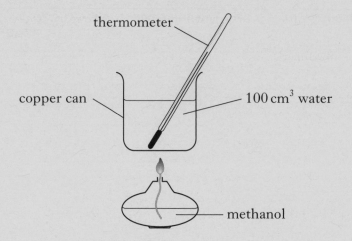

thermometer

copper can

100 cm³ water

methanol

(a) In an experiment, the following results were obtained.

> mass of methanol burned = 0·45 g
>
> temperature rise of water = 10·0 °C

Use these results to calculate the enthalpy of combustion, in kJ mol⁻¹, of methanol.

Show your working clearly.

3

Marks

4. **(continued)**

(*b*) Suggest **two** reasons why the experimental value for the enthalpy of combustion of methanol is different from the value given on page 9 of the data booklet.

2

(5)

[Turn over

Marks

5. Glycerol trinitrate is an explosive with the structure shown.

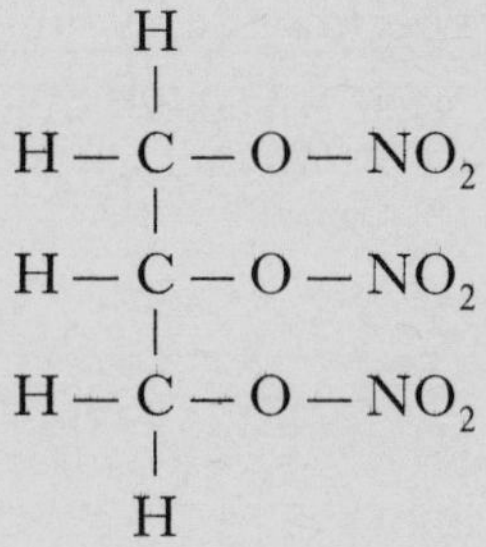

$$
\begin{array}{c}
H \\
| \\
H - C - O - NO_2 \\
| \\
H - C - O - NO_2 \\
| \\
H - C - O - NO_2 \\
| \\
H
\end{array}
$$

(a) Glycerol trinitrate is produced from glycerol.

(i) Draw a structural formula for glycerol.

1

(ii) Name a group of naturally occurring esters that can be hydrolysed to obtain glycerol.

1

(b) When exploded, glycerol trinitrate decomposes to give nitrogen, water, carbon dioxide and oxygen.

Balance the equation for this reaction.

$$C_3H_5N_3O_9(\ell) \rightarrow N_2(g) + H_2O(g) + CO_2(g) + O_2(g)$$

1

(3)

Marks

6. An enzyme found in potatoes can catalyse the decomposition of hydrogen peroxide.

$$2H_2O_2(aq) \rightarrow 2H_2O(\ell) + O_2(g)$$

The rate of the decomposition of hydrogen peroxide can be studied using the apparatus shown.

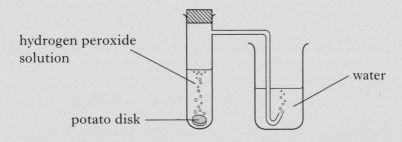

hydrogen peroxide
solution

water

potato disk

(a) Describe how this apparatus can be used to investigate the effect of temperature on the rate of decomposition of hydrogen peroxide.

2

(b) The graph shows how the rate of the enzyme catalysed reaction changes with temperature.

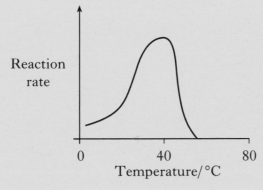

Reaction
rate

0   40   80
Temperature/°C

Why does the reaction rate decrease above the optimum temperature of 40 °C?

1

(3)

Marks

7. An industrial method for the production of ethanol is outlined in the flow diagram.

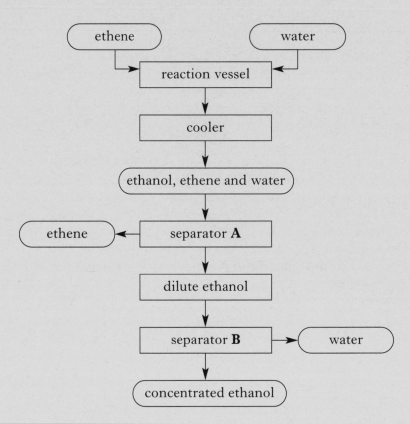

(*a*) The starting materials are ethene and water. Water is a raw material but ethene is not. Why is ethene **not** a raw material?

1

(*b*) (i) Unreacted ethene is removed in separator **A**.

Suggest how the separated ethene could be used to increase the efficiency of the overall process.

1

(ii) Name the process that takes place in separator **B**.

1

Marks

7. **(continued)**

(c) In the reaction vessel, ethanol is produced in an exothermic reaction.

$$C_2H_4(g) \quad + \quad H_2O(g) \quad \rightleftharpoons \quad C_2H_5OH(g)$$

(i) Name the type of chemical reaction that takes place in the reaction vessel.

1

(ii) What would happen to the equilibrium position if the temperature inside the reaction vessel was increased?

1

(iii) If 1·64 kg of ethanol (relative formula mass = 46) is produced from 10·0 kg of ethene (relative formula mass = 28), calculate the percentage yield of ethanol.

2

(7)

Marks

8. "Self-test" kits can be used to check the quantity of alcohol present in a person's breath.

The person blows through a glass tube until a plastic bag at the end is completely filled.

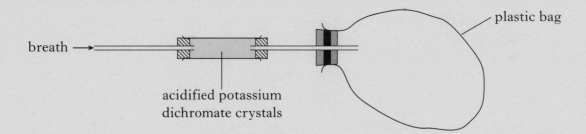

breath

acidified potassium
dichromate crystals

plastic bag

The tube contains orange acidified potassium dichromate crystals that turn green when they react with ethanol. The chemical reaction causing the colour change is:

$$Cr_2O_7^{2-} + 14H^+ + 6e^- \longrightarrow 2Cr^{3+} + 7H_2O$$

orange   green

The more ethanol present in the person's breath, the further along the tube the green colour travels.

(*a*) What is the purpose of the plastic bag?

1

(*b*) Why are the potassium dichromate crystals acidified?

1

(*c*) Name a carbon compound formed by the reaction of ethanol with acidified potassium dichromate crystals.

1

(3)

Marks

9. Car tyres can be made from natural or synthetic rubber.

Natural rubber is a thermoplastic polymer. The monomer used for the polymerisation is called 2-methylbuta-1,3-diene.

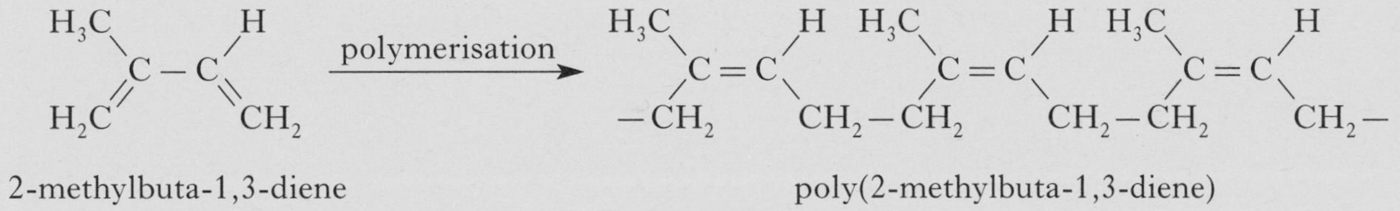

2-methylbuta-1,3-diene   poly(2-methylbuta-1,3-diene)

(*a*) Synthetic rubber is a similar polymer.

Part of the structure of synthetic rubber showing three monomer units linked together is shown.

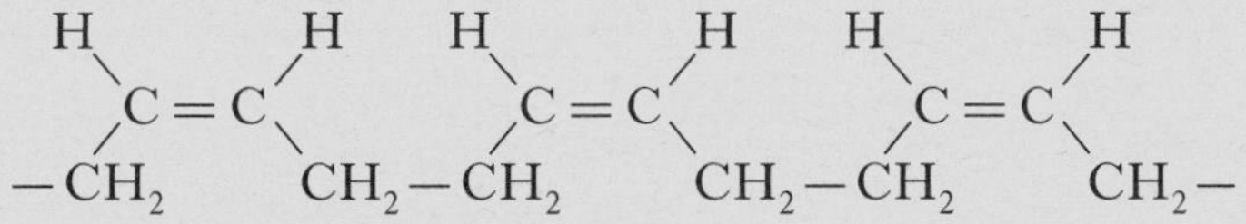

(i) Draw the monomer used to form synthetic rubber.

1

(ii) Name the type of polymerisation involved in the manufacture of synthetic rubber.

1

(*b*) In making tyre rubber, the polymer chains are cross-linked to give a three-dimensional structure by heating with sulphur and sulphur compounds.

In what way will the **properties** of the rubber be improved by cross-linking?

1

(3)

Marks

10. When copper is added to an acid, what happens depends on the acid and the conditions.

(*a*) When copper reacts with dilute nitric acid, nitrate ions are reduced.

Complete the ion-electron equation for the reduction of the nitrate ions.

$$NO_3^-(aq) \longrightarrow NO(g)$$

1

(*b*) Copper reacts with hot, concentrated sulphuric acid to produce sulphur dioxide.

$$Cu + 2H_2SO_4 \rightarrow CuSO_4 + SO_2 + 2H_2O$$

Calculate the volume, in litres, of sulphur dioxide gas that would be produced when 10 g of copper reacts with excess concentrated sulphuric acid.

(Take the molar volume of sulphur dioxide to be 24 litres mol^{-1}.)

Show your working clearly.

2

(3)

Marks

11. (*a*) State the pH of $0 \cdot 01$ mol l^{-1} hydrochloric acid.

1

(*b*) The pH values of $0 \cdot 01$ mol l^{-1} ethanoic acid and $0 \cdot 01$ mol l^{-1} sulphuric acid were measured.

The results were compared to the pH of $0 \cdot 01$ mol l^{-1} hydrochloric acid.

Circle the words in the table that show the correct comparisons.

Solution	pH compared to hydrochloric acid
ethanoic acid	higher / lower / the same
sulphuric acid	higher / lower / the same

2

(*c*) In pure ethanoic acid, the molecules are held together in pairs.

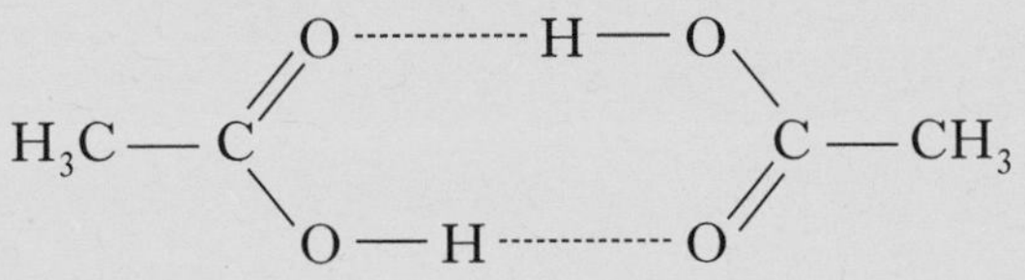

Explain why this happens.

In your answer you should name the type of intermolecular forces involved in pure ethanoic acid and explain how they arise.

2

(5)

[Turn over

Marks

12. The Swiss-born Russian chemist Germain Henri Hess first put forward his law in 1840.

(*a*) (i) Hess's Law can be verified using the reactions summarised below.

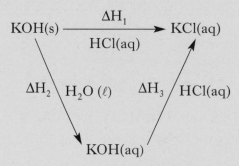

Complete the list of measurements that would have to be made in order to determine ΔH_3.

1   volume of potassium hydroxide solution

2

3

4

5

2

(ii) Hess's Law can be used to obtain enthalpy changes for reactions that cannot be measured directly.

Use the following enthalpy changes

$$KClO_3(s) \ + \ 3Mg(s) \rightarrow KCl(s) + 3MgO(s) \quad \Delta H \ = - \ 1852 \, kJ \, mol^{-1}$$

$$K(s) \ + \ \tfrac{1}{2}Cl_2(g) \ \rightarrow KCl(s) \qquad\qquad\qquad \Delta H \ = - \ 437 \, kJ \, mol^{-1}$$

$$Mg(s) \ + \ \tfrac{1}{2}O_2(g) \ \rightarrow MgO(s) \qquad\qquad\quad \Delta H \ = - \ 602 \, kJ \, mol^{-1}$$

to calculate the enthalpy change in $kJ \, mol^{-1}$, for the reaction:

$$K(s) \ + \ \tfrac{1}{2}Cl_2(g) \ + \ 1\tfrac{1}{2}O_2(g) \ \rightarrow \ KClO_3(s)$$

2

Marks

12. (continued)

(b) The enthalpy change for the reaction

$$K(s) + \tfrac{1}{2}Cl_2(g) + 1\tfrac{1}{2}O_2(g) \rightarrow KClO_3(s)$$

is an example of an enthalpy of formation.

The enthalpy of formation of a compound can be defined as the enthalpy change for the formation of one mole of a compound from its elements as they exist at room temperature.

Write the equation, including state symbols, corresponding to the enthalpy of formation of sodium oxide (Na_2O).

1

(5)

[Turn over

Marks

13. When a mixture of solid sodium hydroxide and solid sodium ethanoate is heated, methane gas and solid sodium carbonate are produced.

$$NaOH(s) + CH_3COONa(s) \rightarrow CH_4(g) + Na_2CO_3(s)$$

(a) Draw a diagram of an apparatus which could be used for this reaction showing how the methane gas can be collected and its volume measured.

2

(b) Name the gas produced when sodium propanoate is heated with solid sodium hydroxide.

1

(3)

Marks

14. Americium-241 is an alpha-emitting radioisotope that is used in a popular type of smoke detector.

   (*a*)   (i) Write a balanced nuclear equation for the alpha decay of americium-241 (atomic number 95).

1

   (ii) Suggest a reason why this type of smoke detector is not regarded as a health hazard, even though it contains a radioactive source.

1

   (*b*)   The americium-241 is present in the smoke detector as the metal oxide.

   If the smoke detector contains $0.00025\,g$ of americium-241 oxide, AmO_2, calculate the mass present, in grams, of americium-241.

1

(3)

[Turn over

Marks

15. Chlorine can be produced commercially from concentrated sodium chloride solution in a membrane cell.

Only sodium ions can pass through the membrane. These ions move in the direction shown in the diagram.

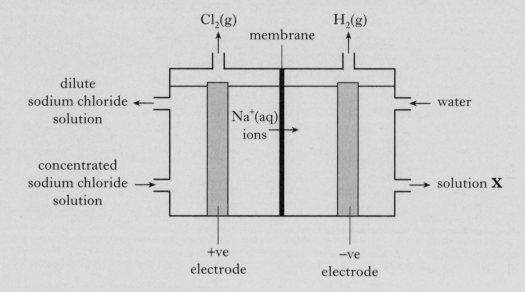

The reactions at each electrode are:

+ve electrode: $2Cl^-(aq) \rightarrow Cl_2(g) + 2e^-$

−ve electrode: $2H_2O(\ell) + 2e^- \rightarrow H_2(g) + 2OH^-(aq)$

(a) Write the overall redox equation for the reaction in the membrane cell.

1

(b)   (i)   Name solution **X**.

1

(ii)   Hydrogen gas is produced at the negative electrode in the membrane cell.

Suggest why this gas could be a valuable resource in the future.

1

DO NOT WRITE IN THIS MARGIN

Marks

15. (continued)

(*c*) Calculate the mass of chlorine, in kilograms, produced in a membrane cell using a current of 80 000 A for 10 hours.

Show your working clearly.

3

(6)

[Turn over

Marks

16. Infrared spectroscopy can be used to help identify the bonds which are present in an organic molecule.

    Different bonds absorb infrared radiation of different wave numbers.

    The table below shows the range of wave numbers of infrared radiation absorbed by the bonds indicated with thicker lines.

Bond	Wave number range/cm^{-1}
O — H	3650 – 3590
C ≡ C — H	3300
C — C — H	2962 – 2853
C ≡ C	2260 – 2100
C — O	1150 – 1070

The infrared spectrum and full structural formula for compound ① are shown.

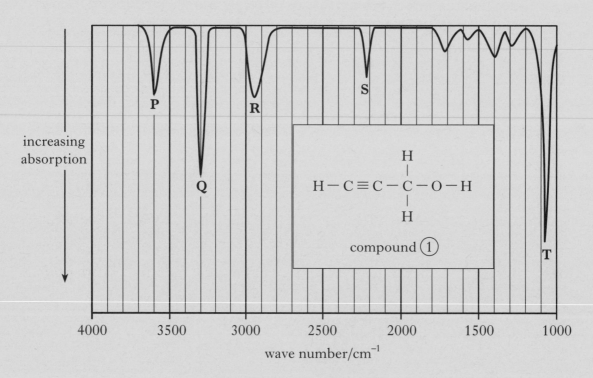

(a) In the above full structural formula for compound ①, use an arrow to indicate the bond that could be responsible for the absorption at **T** in its infrared spectrum.

1

DO NOT
WRITE IN
THIS
MARGIN

Marks

16. (continued)

(*b*) A series of reactions is carried out starting with compound ①.

$$\text{compound ①} \xrightarrow{\text{1 mol } H_2} \text{compound ②} \xrightarrow{\text{1 mol } H_2} \text{compound ③}$$

(i) Give the letters for the **two** absorptions which would **not** appear in the infrared spectrum for compound ②.

1

(ii) Name compound ③.

1

(3)

[END OF QUESTION PAPER]

ADDITIONAL SPACE FOR ANSWERS

ADDITIONAL SPACE FOR ANSWERS

ADDITIONAL SPACE FOR ANSWERS

ADDITIONAL SPACE FOR ANSWERS

ADDITIONAL SPACE FOR ANSWERS

HIGHER
2007

[BLANK PAGE]

FOR OFFICIAL USE

Total
Section B

X012/301

NATIONAL
QUALIFICATIONS
2007

TUESDAY, 29 MAY
9.00 AM – 11.30 AM

CHEMISTRY
HIGHER

Fill in these boxes and read what is printed below.

Full name of centre

Town

Forename(s)

Surname

Date of birth

Day Month Year

Scottish candidate number

Number of seat

Reference may be made to the Chemistry Higher and Advanced Higher Data Booklet .

SECTION A—Questions 1–40 (40 marks)

Instructions for completion of **Section A** are given on page two.

For this section of the examination you must use an **HB pencil**.

SECTION B (60 marks)

1   All questions should be attempted.

2   The questions may be answered in any order but all answers are to be written in the spaces provided in this answer book, **and must be written clearly and legibly in ink**.

3   Rough work, if any should be necessary, should be written in this book and then scored through when the fair copy has been written. If further space is required, a supplementary sheet for rough work may be obtained from the invigilator.

4   Additional space for answers will be found at the end of the book. If further space is required, supplementary sheets may be obtained from the invigilator and should be inserted inside the **front** cover of this book.

5   The size of the space provided for an answer should not be taken as an indication of how much to write. It is not necessary to use all the space.

6   Before leaving the examination room you must give this book to the invigilator. If you do not, you may lose all the marks for this paper.

SCOTTISH
QUALIFICATIONS
AUTHORITY

SECTION A

Read carefully

1 Check that the answer sheet provided is for **Chemistry Higher (Section A)**.

2 For this section of the examination you must use an **HB pencil** and, where necessary, an eraser.

3 Check that the answer sheet you have been given has **your name**, **date of birth**, **SCN** (Scottish Candidate Number) and **Centre Name** printed on it.

Do not change any of these details.

4 If any of this information is wrong, tell the Invigilator immediately.

5 If this information is correct, **print** your name and seat number in the boxes provided.

6 The answer to each question is **either** A, B, C or D. Decide what your answer is, then, using your pencil, put a horizontal line in the space provided (see sample question below).

7 There is **only one correct** answer to each question.

8 Any rough working should be done on the question paper or the rough working sheet, **not** on your answer sheet.

9 At the end of the exam, put the **answer sheet for Section A inside the front cover of your answer book**.

Sample Question

To show that the ink in a ball-pen consists of a mixture of dyes, the method of separation would be

      A chromatography

      B fractional distillation

      C fractional crystallisation

      D filtration.

The correct answer is **A**—chromatography. The answer **A** has been clearly marked in **pencil** with a horizontal line (see below).

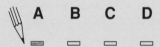

Changing an answer

If you decide to change your answer, carefully erase your first answer and using your pencil, fill in the answer you want. The answer below has been changed to **D**.

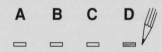

1. Which of the following compounds contains **both** a halide ion and a transition metal ion?

   A   Iron oxide

   B   Silver bromide

   C   Potassium permanganate

   D   Copper iodate

2. Which of the following substances is a non-conductor but becomes a good conductor on melting?

   A   Solid potassium fluoride

   B   Solid argon

   C   Solid potassium

   D   Solid tetrachloromethane

3. Particles with the same electron arrangement are said to be isoelectronic.

   Which of the following compounds contains ions which are isoelectronic?

   A Na_2S

   B $MgCl_2$

   C   KBr

   D $CaCl_2$

4. The graph shows the variation of concentration of a reactant with time as a reaction proceeds.

   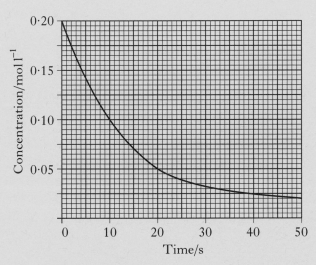

   What is the average reaction rate during the first 20 s?

   A $0.0025 \ mol \, l^{-1} s^{-1}$

   B $0.0050 \ mol \, l^{-1} s^{-1}$

   C $0.0075 \ mol \, l^{-1} s^{-1}$

   D $0.0150 \ mol \, l^{-1} s^{-1}$

5. The potential energy diagram below refers to the reversible reaction involving reactants **R** and products **P**.

   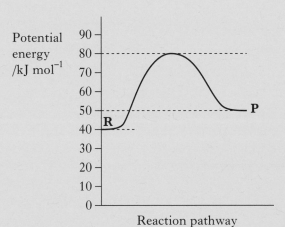

   What is the enthalpy change, in $kJ \, mol^{-1}$, for the reverse reaction $\mathbf{P} \rightarrow \mathbf{R}$?

   A   + 30

   B   + 10

   C   − 10

   D   − 40

6. The enthalpy of neutralisation in an acid/alkali reaction is **always** the energy released in

   A   the formation of one mole of salt

   B   the formation of one mole of water

   C   the neutralisation of one mole of acid

   D   the neutralisation of one mole of alkali.

7. Which equation represents the first ionisation energy of a diatomic element, X_2?

   A $\frac{1}{2}X_2(s) \rightarrow X^+(g)$

   B $\frac{1}{2}X_2(g) \rightarrow X^-(g)$

   C $X(g) \rightarrow X^+(g)$

   D $X(s) \rightarrow X^-(g)$

8. Which of the following chlorides is likely to have **least** ionic character?

   A $BeCl_2$

   B $CaCl_2$

   C   LiCl

   D   CsCl

[Turn over

9. Which of the following chlorides is most likely to be soluble in tetrachloromethane, CCl_4?

   A Barium chloride

   B Caesium chloride

   C Calcium chloride

   D Phosphorus chloride

10. Which of the following compounds exists as discrete molecules?

   A Sulphur dioxide

   B Silicon dioxide

   C Aluminium oxide

   D Iron(II) oxide

11. An element (melting point above 3000 °C) forms an oxide which is a gas at room temperature.

   Which type of bonding is likely to be present in the element?

   A Metallic

   B Polar covalent

   C Non-polar covalent

   D Ionic

12. Which of the following compounds has polar molecules?

   A CO_2

   B NH_3

   C CCl_4

   D CH_4

13. How many moles of oxygen atoms are in 0·5 mol of carbon dioxide?

   A 0·25

   B 0·5

   C 1

   D 2

14. A fullerene molecule consists of 60 carbon atoms.

   Approximately how many such molecules are present in 12 g of this type of carbon?

   A $1·0 \times 10^{22}$

   B $1·2 \times 10^{23}$

   C $6·0 \times 10^{23}$

   D $3·6 \times 10^{25}$

15. Avogadro's Constant is the same as the number of

   A molecules in 16·0 g of oxygen

   B atoms in 20·2 g of neon

   C formula units in 20·0 g of sodium hydroxide

   D ions in 58·5 g of sodium chloride.

16. The equation for the complete combustion of propane is:

   $$C_3H_8(g) + 5O_2(g) \rightarrow 3CO_2(g) + 4H_2O(\ell)$$

   $30 \, cm^3$ of propane is mixed with $200 \, cm^3$ of oxygen and the mixture is ignited.

   What is the volume of the resulting gas mixture? (All volumes are measured at the same temperature and pressure.)

   A $90 \, cm^3$

   B $120 \, cm^3$

   C $140 \, cm^3$

   D $210 \, cm^3$

17. A mixture of carbon monoxide and hydrogen can be converted into water and a mixture of hydrocarbons.

   $$nCO + (2n + 1)H_2 \rightarrow nH_2O + \text{hydrocarbons}$$

   What is the general formula for the hydrocarbons produced?

   A C_nH_{2n-2}

   B C_nH_{2n}

   C C_nH_{2n+1}

   D C_nH_{2n+2}

18. Chemical processes are used to produce a petrol that burns more efficiently.

    Which of the following types of hydrocarbon does **not** improve the burning efficiency of petrol?

    A straight-chain alkanes

    B branched-chain alkanes

    C cycloalkanes

    D aromatics

19. Which of the following is an aldehyde?

20. The dehydration of butan-2-ol can produce two isomeric alkenes, but-1-ene and but-2-ene.

    Which of the following alkanols can similarly produce, on dehydration, a pair of isomeric alkenes?

    A Propan-2-ol

    B Pentan-3-ol

    C Hexan-3-ol

    D Heptan-4-ol

21. Which of the following reactions can be classified as reduction?

    A $CH_3CH_2OH \rightarrow CH_3COOH$

    B $CH_3CH(OH)CH_3 \rightarrow CH_3COCH_3$

    C $CH_3CH_2COCH_3 \rightarrow CH_3CH_2CH(OH)CH_3$

    D $CH_3CH_2CHO \rightarrow CH_3CH_2COOH$

22. Which of the following compounds would react with sodium hydroxide solution to form a salt?

    A CH_3CHO

    B CH_3COOH

    C CH_3COCH_3

    D CH_3CH_2OH

23. The extensive use of which type of compound is thought to contribute significantly to the depletion of the ozone layer?

    A Oxides of carbon

    B Hydrocarbons

    C Oxides of sulphur

    D Chlorofluorocarbons

24. Propene is used in the manufacture of addition polymers.

    What type of reaction is used to produce propene from propane?

    A Addition

    B Cracking

    C Hydrogenation

    D Oxidation

25. Cured polyester resins

    A are used as textile fibres

    B are long chain molecules

    C are formed by addition polymerisation

    D have a three-dimensional structure with cross linking.

    [Turn over

26. What is the structural formula for glycerol?

A CH$_2$OH
   |
  CH$_2$
   |
  CH$_2$OH

B CH$_2$OH
   |
  CH$_2$OH

C CH$_2$OH
   |
  CHOH
   |
  CH$_2$COOH

D CH$_2$OH
   |
  CHOH
   |
  CH$_2$OH

27. The monomer units used to construct enzyme molecules are

A alcohols

B esters

C amino acids

D fatty acids.

28. In α-amino acids the amino group is on the carbon atom adjacent to the acid group.

Which of the following is an α-amino acid?

A CH$_3$ — CH — COOH
         |
       CH$_2$ — NH$_2$

B CH$_2$ — CH — COOH
   | |
  SH   NH$_2$

C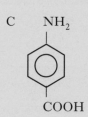

D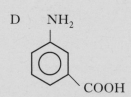

29. Which of the following compounds is **not** a raw material in the chemical industry?

A Benzene

B Water

C Iron ore

D Sodium chloride

30. $N_2(g) + 2O_2(g) \rightarrow 2NO_2(g)$ $\Delta H = +88\,kJ$
$N_2(g) + 2O_2(g) \rightarrow N_2O_4(g)$ $\Delta H = +10\,kJ$

The enthalpy change for the reaction

$$2NO_2(g) \rightarrow N_2O_4(g)$$

will be

A +98 kJ

B +78 kJ

C −78 kJ

D −98 kJ.

31. A catalyst is used in the Haber Process.

$$N_2(g) + 3H_2(g) \rightleftharpoons 2NH_3(g)$$

Which of the following best describes the action of the catalyst?

A Increases the rate of the forward reaction only

B Increases the rate of the reverse reaction only

C Increases the rate of both the forward and reverse reactions

D Changes the position of the equilibrium of the reaction

32. In which of the following systems will the equilibrium be **unaffected** by a change in pressure?

A $2NO_2(g) \rightleftharpoons N_2O_4(g)$

B $H_2(g) + I_2(g) \rightleftharpoons 2HI(g)$

C $N_2(g) + 3H_2(g) \rightleftharpoons 2NH_3(g)$

D $2NO(g) + O_2(g) \rightleftharpoons 2NO_2(g)$

33. On the structure shown, four hydrogen atoms have been replaced by letters **W**, **X**, **Y** and **Z**.

Which letter corresponds to the hydrogen atom which can ionise most easily in aqueous solution?

A **W**

B **X**

C **Y**

D **Z**

34. The concentration of $OH^-(aq)$ ions in a solution is $0 \cdot 1 \text{ mol l}^{-1}$.

What is the pH of the solution?

A 1

B 8

C 13

D 14

35. A lemon juice is found to have a pH of 3 and an apple juice a pH of 5.

From this information, the concentrations of $H^+(aq)$ ions in the lemon juice and apple juice are in the proportion (ratio)

A 100 : 1

B 1 : 100

C 20 : 1

D 3 : 5.

36. Which line in the table is correct for $0 \cdot 1 \text{ mol l}^{-1}$ sodium hydroxide solution compared with $0 \cdot 1 \text{ mol l}^{-1}$ ammonia solution?

	pH	Conductivity
A	higher	lower
B	higher	higher
C	lower	higher
D	lower	lower

37. The iodate ion, IO_3^-, can be converted to iodine.

Which is the correct ion-electron equation for the reaction?

A $2IO_3^-(aq) + 12H^+(aq) + 12e^- \rightarrow 2I^-(aq) + 6H_2O(\ell)$

B $IO_3^-(aq) + 6H^+(aq) + 7e^- \rightarrow I^-(aq) + 3H_2O(\ell)$

C $2IO_3^-(aq) + 12H^+(aq) + 11e^- \rightarrow I_2(aq) + 6H_2O(\ell)$

D $2IO_3^-(aq) + 12H^+(aq) + 10e^- \rightarrow I_2(aq) + 6H_2O(\ell)$

38. Which of the following is a redox reaction?

A $Mg + 2HCl \rightarrow MgCl_2 + H_2$

B $MgO + 2HCl \rightarrow MgCl_2 + H_2O$

C $MgCO_3 + 2HCl \rightarrow MgCl_2 + H_2O + CO_2$

D $Mg(OH)_2 + 2HCl \rightarrow MgCl_2 + 2H_2O$

39. Strontium-90 is a radioisotope.

What is the neutron to proton ratio in an atom of this isotope?

A $0 \cdot 730$

B $1 \cdot 00$

C $1 \cdot 37$

D $2 \cdot 37$

40. The diagram shows the paths of alpha, beta and gamma radiations as they pass through an electric field.

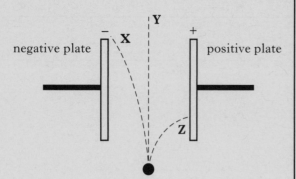

radioactive source

Which line in the table correctly identifies the types of radiation which follow paths **X**, **Y** and **Z**?

	Path X	Path Y	Path Z
A	gamma	beta	alpha
B	beta	gamma	alpha
C	beta	alpha	gamma
D	alpha	gamma	beta

Candidates are reminded that the answer sheet MUST be returned INSIDE the front cover of this answer book.

DO NOT
WRITE IN
THIS
MARGIN

Marks

SECTION B

All answers must be written clearly and legibly in ink.

1. (*a*) Atoms of different elements have different attractions for bonded electrons.

What term is used as a measure of the attraction an atom involved in a bond has for the electrons of the bond?

1

(*b*) Atoms of different elements are different sizes.

What is the trend in atomic size across the period from sodium to argon?

1

(*c*) Atoms of different elements have different ionisation energies.

Explain clearly why the first ionisation energy of potassium is less than the first ionisation energy of sodium.

2

(4)

[Turn over

Marks

2. Carbon compounds take part in a wide variety of chemical reactions.

(a)

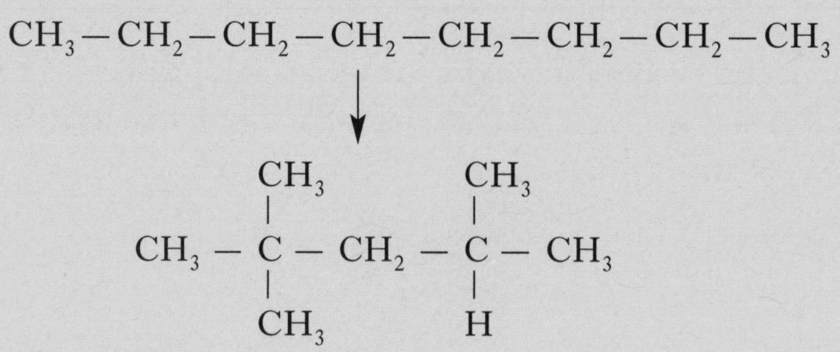

$$CH_3-CH_2-CH_2-CH_2-CH_2-CH_2-CH_2-CH_3$$

$$\downarrow$$

$$\underset{\underset{CH_3}{|}}{\overset{\overset{CH_3}{|}}{CH_3-C}}-CH_2-\underset{\underset{H}{|}}{\overset{\overset{CH_3}{|}}{C}}-CH_3$$

Name this type of chemical reaction.

1

(b)

$$C_3H_6O \xrightarrow{\text{oxidation}} \text{propanoic acid}$$

Draw a structural formula for C_3H_6O.

1

(c) Kevlar is an aromatic polyamide made by condensation polymerisation.
Give **one** use for Kevlar.

1

(3)

Marks

3. Tritium, $_1^3$H, is an isotope of hydrogen. It is formed in the upper atmosphere when neutrons from cosmic rays are captured by nitrogen atoms.

$$_7^{14}N \ + \ _0^1n \ \rightarrow \ _6^{12}C \ + \ _1^3H$$

Tritium atoms then decay by beta-emission.

$$_1^3H \ \rightarrow \qquad +$$

(*a*) Complete the nuclear equation above for the beta-decay of tritium atoms.

1

(*b*) In the upper atmosphere, tritium atoms are present in some water molecules. Over the years, the concentration of tritium atoms in rain has remained fairly constant.

(i) Why does the concentration of tritium in rain remain fairly constant?

1

(ii) The concentration of tritium atoms in fallen rainwater is found to decrease over time. The age of any product made with water can be estimated by measuring the concentration of tritium atoms.

In a bottle of wine, the concentration of tritium atoms was found to be $\frac{1}{8}$ of the concentration found in rain.

Given that the half-life of tritium is 12·3 years, how old is the wine?

1

(3)

[Turn over

Marks

4. Hydrogen gas is widely regarded as a very valuable fuel for the future.

(*a*) Hydrogen can be produced from methane by steam reforming. The process proceeds in two steps.

Step 1: $CH_4(g)$ + $H_2O(g)$ → $CO(g)$ + $3H_2(g)$

Step 2: $CO(g)$ + $H_2O(g)$ → $CO_2(g)$ + $H_2(g)$

(i) What name is given to the gas mixture produced in step 1?

1

(ii) Using this process, how many moles of hydrogen gas can be produced overall from one mole of methane?

1

(*b*) Hydrogen can be produced in the lab from dilute sulphuric acid. The apparatus shown below can be used to investigate the quantity of electrical charge required to form one mole of hydrogen gas.

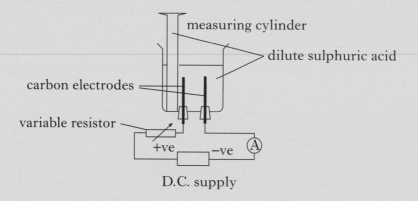

(i) Above which electrode should the measuring cylinder be placed to collect the hydrogen gas?

1

(ii) In addition to the current, what **two** measurements should be taken?

1

Marks

5. The energy changes taking place during chemical reactions have many everyday uses.

   (*a*) Some portable cold packs make use of the temperature drop that takes place when the chemicals in the pack dissolve in water.

   Name the type of reaction that results in a fall in temperature.

   1

   (*b*) Flameless heaters are used by mountain climbers to heat food and drinks. The chemical reaction in a flameless heater releases 45 kJ of energy.

   If 200 g of water is heated using this heater, calculate the rise in temperature of the water, in °C.

   1
   (2)

   [Turn over

Marks

6. Temperature has a very significant effect on the rate of a chemical reaction.

(a) The reaction shown below can be used to investigate the effect of temperature on reaction rate.

$$5(COOH)_2(aq) + 6H^+(aq) + 2MnO_4^-(aq) \rightarrow 2Mn^{2+}(aq) + 10CO_2(g) + 8H_2O(\ell)$$

The instructions for such an investigation are shown below.

Procedure

1. Using syringes, add $5\,cm^3$ of sulphuric acid, $2\,cm^3$ of potassium permanganate solution and $40\,cm^3$ of water to a $100\,cm^3$ dry glass beaker.

2. Heat the mixture to about 40 °C.

3. Place the beaker on a white tile and measure $1\,cm^3$ of oxalic acid solution into a syringe.

4. Add the oxalic acid to the mixture in the beaker as quickly as possible and at the same time start the timer.

5. Gently stir the reaction mixture with the thermometer.

6. When the reaction is over, stop the timer and record the time. Measure and record the temperature of the reaction mixture.

7. Repeat the experiment three times but heat the initial sulphuric acid/potassium permanganate/water mixtures first to 50 °C, then to 60 °C and finally to 70 °C.

(i) What colour change indicates that the reaction is over?

1

(ii) With each of the experiments, the temperature of the solution was measured both during heating and at the end of the reaction.

When plotting graphs of the reaction rate against temperature, it is the temperature measured at the end of reaction, rather than the temperature measured while heating, that is used.

Give a reason for this.

1

Marks

6. (continued)

(*b*) The graph shows the distribution of kinetic energy for molecules in a reaction mixture at a given temperature.

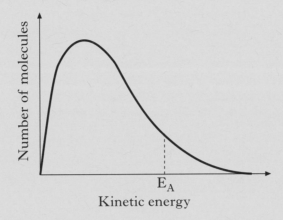

Why does a small increase in temperature produce a large increase in reaction rate?

1

(3)

[Turn over

Marks

7. Magnesium metal can be extracted from sea water.

An outline of the reactions involved is shown in the flow diagram.

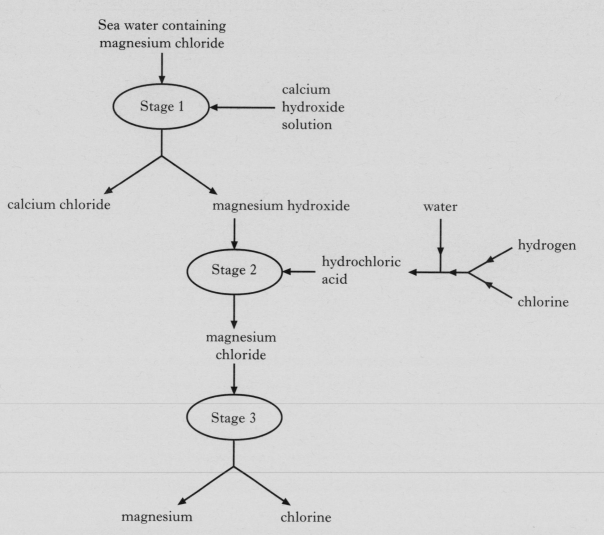

(a) Why can the magnesium hydroxide be easily separated from the calcium chloride at Stage 1?

1

(b) Name the type of chemical reaction taking place at Stage 2.

1

Marks

7. **(continued)**

(c)   Give **two** different features of this process that make it economical.

2

(d)   At Stage 3, electrolysis of molten magnesium chloride takes place.

If a current of 200 000 A is used, calculate the mass of magnesium, in kg, produced in 1 minute.

Show your working clearly.

3

(7)

[Turn over

Marks

8. One of the chemicals released in a bee sting is an ester that has the structure shown.

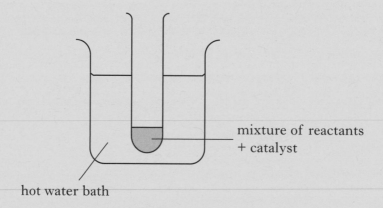

This ester can be produced by the reaction of an alcohol with an alkanoic acid.

(*a*) Name this acid.

1

(*b*) The ester can be prepared in the lab by heating a mixture of the reactants with a catalyst.

mixture of reactants
+ catalyst

hot water bath

(i) Name the catalyst used in the reaction.

1

(ii) What improvement could be made to the experimental set-up shown in the diagram?

1

Marks

8. (continued)

(*c*) If there is a 65% yield, calculate the mass of ester produced, in grams, when 4·0 g of the alcohol reacts with a slight excess of the acid.

(Mass of one mole of the alcohol = 88 g;
mass of one mole of the ester = 130 g)

Show your working clearly.

2

(5)

[Turn over

DO NOT WRITE IN THIS MARGIN

Marks

9. Phenylalanine and alanine are both amino acids.

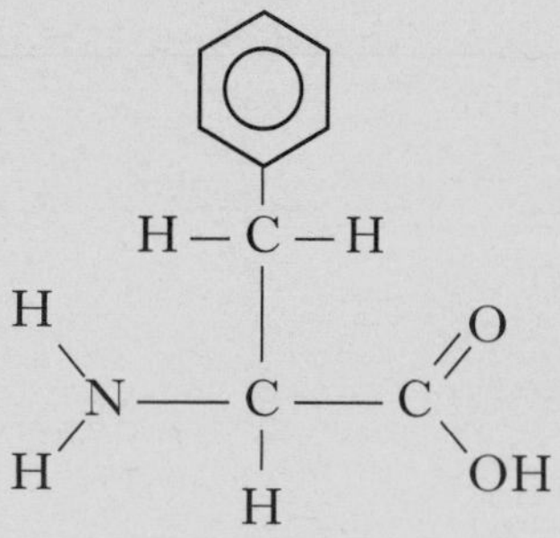

phenylalanine

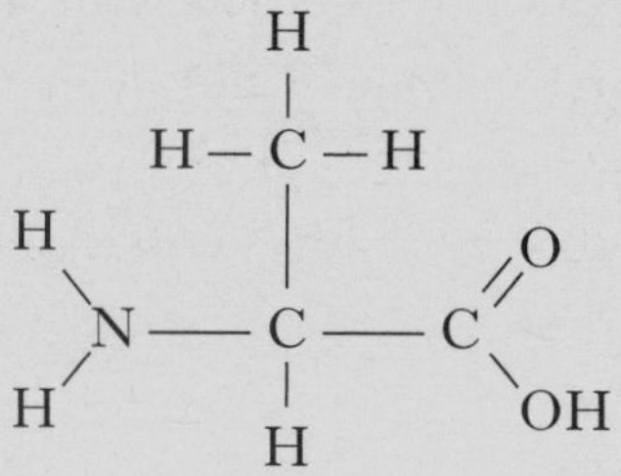

alanine

(*a*) Phenylalanine is an essential amino acid.

(i) What is meant by an essential amino acid? **1**

(ii) How many hydrogen atoms are present in a molecule of phenylalanine? **1**

(*b*) Phenylalanine and alanine can react to form the dipeptide shown.

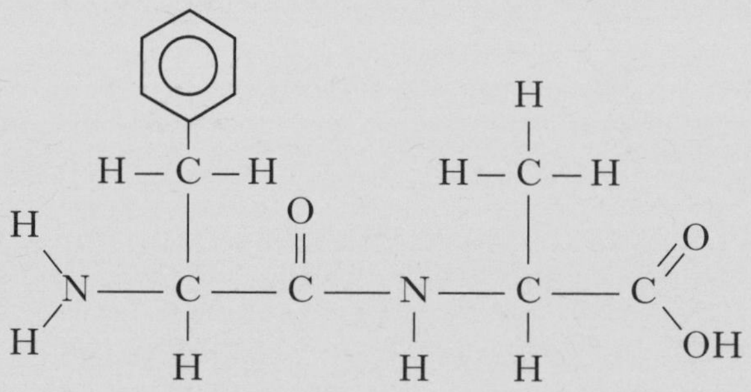

Circle the peptide link in this molecule. **1**

(*c*) Draw a structural formula for the other dipeptide that can be formed from phenylalanine and alanine. **1**

(4)

Marks

10. A student carried out three experiments involving the reaction of excess magnesium ribbon with dilute acids. The rate of hydrogen production was measured in each of the three experiments.

Experiment	Acid
1	$100 \, cm^3$ of $0.10 \, mol \, l^{-1}$ sulphuric acid
2	$50 \, cm^3$ of $0.20 \, mol \, l^{-1}$ sulphuric acid
3	$100 \, cm^3$ of $0.10 \, mol \, l^{-1}$ hydrochloric acid

The equation for **Experiment 1** is shown.

$$Mg(s) \;+\; H_2SO_4(aq) \;\rightarrow\; MgSO_4(aq) \;+\; H_2(g)$$

(a) The curve obtained for **Experiment 1** is drawn on the graph.

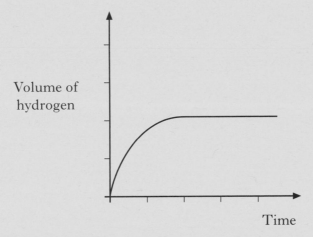

Draw curves on the graph to show the results obtained for **Experiment 2** and **Experiment 3**.

Label each curve clearly.

(An additional graph, if required, can be found on *Page thirty*.)   2

(b) The mass of magnesium used in **Experiment 1** was $0.50 \, g$.

For this experiment, calculate the mass of magnesium, in grams, left unreacted.

2

(4)

Marks

11. Nitrogen dioxide gas can be prepared in different ways.

(*a*) It is manufactured industrially as part of the Ostwald process. In the first stage of the process, nitrogen monoxide is produced by passing ammonia and oxygen over a platinum catalyst.

$$NH_3(g) \quad + \quad O_2(g) \quad \rightarrow \quad NO(g) \quad + \quad H_2O(g)$$

   (i) Balance the above equation.

1

   (ii) Platinum metal is a heterogeneous catalyst for this reaction.

What is meant by a **heterogeneous** catalyst?

1

   (iii) The nitrogen monoxide then combines with oxygen in an exothermic reaction to form nitrogen dioxide.

$$2NO(g) \quad + \quad O_2(g) \quad \rightleftharpoons \quad 2NO_2(g)$$

What happens to the yield of nitrogen dioxide gas if the reaction mixture is cooled?

1

(*b*) In the lab, nitrogen dioxide gas can be prepared by heating copper(II) nitrate.

$$Cu(NO_3)_2(s) \quad \rightarrow \quad CuO(s) \quad + \quad 2NO_2(g) \quad + \quad \tfrac{1}{2}O_2(g)$$

   (i) Calculate the volume of nitrogen dioxide gas produced when 2·0 g of copper(II) nitrate is completely decomposed on heating.

(Take the molar volume of nitrogen dioxide to be 24 litres mol^{-1}.)

Show your working clearly.

2

Marks

11. **(*b*)** **(continued)**

(ii) Nitrogen dioxide has a boiling point of 22 °C.

Complete the diagram to show how nitrogen dioxide can be separated and collected.

copper(II) nitrate

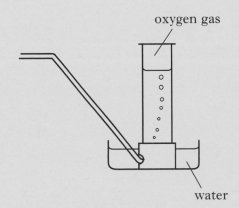

1

(6)

[Turn over

Marks

12. Ethyne is the first member of the homologous series called the alkynes.

(*a*) Ethyne can undergo addition reactions as shown in the flow diagram.

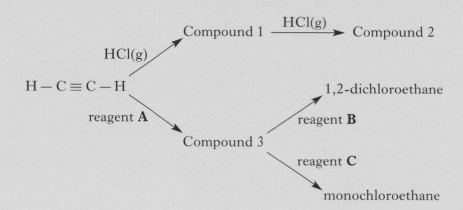

(i) Compound 2 is an isomer of 1,2-dichloroethane.

Draw a structural formula for compound 2.

1

(ii) Reagents **A**, **B** and **C** are three **different** diatomic gases.

Using information in the flow diagram, identify reagents **A**, **B** and **C**.

Reagent	Gas
A	
B	
C	

1

Marks

12. **(continued)**

(*b*) The equation for the enthalpy of formation of ethyne is:

$$2C(s) \quad + \quad H_2(g) \quad \rightarrow \quad C_2H_2(g)$$

Use the enthalpies of combustion of carbon, hydrogen and ethyne given in the data booklet to calculate the enthalpy of formation of ethyne, in $kJ\,mol^{-1}$.

Show your working clearly.

2

(4)

[Turn over

Marks

13. Compared to other gases made up of molecules of similar molecular masses, ammonia has a relatively high boiling point.

ammonia

(a) In terms of the intermolecular bonding present, **explain clearly** why ammonia has a relatively high boiling point.

2

(b) Amines can be produced by reacting ammonia with an aldehyde or a ketone. This reaction, an example of reductive amination, occurs in two stages.

Stage 1 Amination

$$CH_3-C-CH_3 \ + \ NH_3 \longrightarrow CH_3-C-CH_3$$

Stage 2 Reduction

(i) Give another name for the type of reaction taking place in Stage 2.

1

DO NOT WRITE IN THIS MARGIN

Marks

13. (*b*) (continued)

(ii) Draw the structural formula for the amine produced when butanal undergoes reductive amination with ammonia. **1**

(4)

[Turn over

Marks

14. Foodstuffs have labels that list ingredients and provide nutritional information.

(a) The label on a tub of margarine lists **hydrogenated vegetable oils** as one of the ingredients.

Why have some of the vegetable oils in this product been hydrogenated?

1

(b) Potassium sorbate is a salt that is used as a preservative in margarine.

Potassium sorbate dissolves in water to form an alkaline solution.

What does this indicate about sorbic acid?

1

(c) The nutritional information states that 100 g of margarine contains 0·70 g of sodium. The sodium is present as sodium chloride (NaCl).

Calculate the mass of sodium chloride, in g, present in every 100 g of margarine.

1

(3)

Marks

15. Seaweeds are a rich source of iodine in the form of iodide ions. The mass of iodine in a seaweed can be found using the procedure outlined below.

(*a*) **Step 1**

The seaweed is dried in an oven and ground into a fine powder. Hydrogen peroxide solution is then added to oxidise the iodide ions to iodine molecules. The ion-electron equation for the reduction reaction is shown.

$$H_2O_2(aq) + 2H^+(aq) + 2e^- \rightarrow 2H_2O(\ell)$$

Write a balanced redox equation for the reaction of hydrogen peroxide with iodide ions.

1

(*b*) **Step 2**

Using starch solution as an indicator, the iodine solution is then titrated with sodium thiosulphate solution to find the mass of iodine in the sample. The balanced equation for the reaction is shown.

$$2Na_2S_2O_3(aq) + I_2(aq) \rightarrow 2NaI(aq) + Na_2S_4O_6(aq)$$

In an analysis of seaweed, $14.9 \, cm^3$ of $0.00500 \, mol \, l^{-1}$ sodium thiosulphate solution was required to reach the end-point.

Calculate the mass of iodine present in the seaweed sample.

Show your working clearly.

3

(4)

[END OF QUESTION PAPER]

SPACE FOR ANSWERS

ADDITIONAL GRAPH FOR QUESTION 10(*a*)

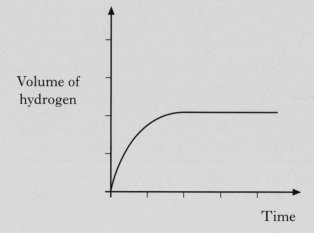

DO NOT WRITE IN THIS MARGIN

ADDITIONAL SPACE FOR ANSWERS

[BLANK PAGE]

[BLANK PAGE]

FOR OFFICIAL USE

Total
Section B

X012/301

NATIONAL
QUALIFICATIONS
2008

FRIDAY, 30 MAY
9.00 AM – 11.30 AM

CHEMISTRY
HIGHER

Fill in these boxes and read what is printed below.

Full name of centre

Town

Forename(s)

Surname

Date of birth

Day Month Year

Scottish candidate number

Number of seat

Reference may be made to the Chemistry Higher and Advanced Higher Data Booklet.

SECTION A—Questions 1–40 (40 marks)

Instructions for completion of **Section A** are given on page two.

For this section of the examination you must use an **HB pencil**.

SECTION B (60 marks)

1   All questions should be attempted.

2   The questions may be answered in any order but all answers are to be written in the spaces provided in this answer book, **and must be written clearly and legibly in ink**.

3   Rough work, if any should be necessary, should be written in this book and then scored through when the fair copy has been written. If further space is required, a supplementary sheet for rough work may be obtained from the invigilator.

4   Additional space for answers will be found at the end of the book. If further space is required, supplementary sheets may be obtained from the invigilator and should be inserted inside the **front** cover of this book.

5   The size of the space provided for an answer should not be taken as an indication of how much to write. It is not necessary to use all the space.

6   Before leaving the examination room you must give this book to the invigilator. If you do not, you may lose all the marks for this paper.

SECTION A

Read carefully

1 Check that the answer sheet provided is for **Chemistry Higher (Section A)**.

2 For this section of the examination you must use an **HB pencil** and, where necessary, an eraser.

3 Check that the answer sheet you have been given has **your name**, **date of birth**, **SCN** (Scottish Candidate Number) and **Centre Name** printed on it.

   Do not change any of these details.

4 If any of this information is wrong, tell the Invigilator immediately.

5 If this information is correct, **print** your name and seat number in the boxes provided.

6 The answer to each question is **either** A, B, C or D. Decide what your answer is, then, using your pencil, put a horizontal line in the space provided (see sample question below).

7 There is **only one correct** answer to each question.

8 Any rough working should be done on the question paper or the rough working sheet, **not** on your answer sheet.

9 At the end of the exam, put the **answer sheet for Section A inside the front cover of your answer book**.

Sample Question

To show that the ink in a ball-pen consists of a mixture of dyes, the method of separation would be

   A chromatography

   B fractional distillation

   C fractional crystallisation

   D filtration.

The correct answer is **A**—chromatography. The answer **A** has been clearly marked in **pencil** with a horizontal line (see below).

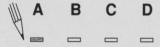

Changing an answer

If you decide to change your answer, carefully erase your first answer and using your pencil, fill in the answer you want. The answer below has been changed to **D**.

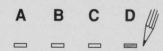

1. Solutions of barium chloride and silver nitrate are mixed together.

   The reaction that takes place is an example of

   A displacement

   B neutralisation

   C oxidation

   D precipitation.

2. Two rods are placed in dilute sulphuric acid as shown.

   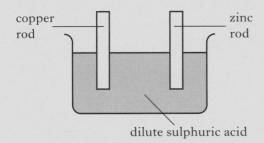

   copper rod   zinc rod

   dilute sulphuric acid

   Which of the following would be observed?

   A No gas is given off.

   B Gas is given off at only the zinc rod.

   C Gas is given off at only the copper rod.

   D Gas is given off at both rods.

3. An element was burned in air. The product was added to water, producing a solution with a pH less than 7.

   The element could be

   A carbon

   B hydrogen

   C sodium

   D tin.

4. A mixture of sodium chloride and sodium sulphate is known to contain 0·6 mol of chloride ions and 0·2 mol of sulphate ions.

   How many moles of sodium ions are present?

   A 0·4

   B 0·5

   C 0·8

   D 1·0

5. The following results were obtained in the reaction between marble chips and dilute hydrochloric acid.

Time/minutes	0	2	4	6	8	10
Total volume of carbon dioxide produced/cm^3	0	52	68	78	82	84

   What is the average rate of production of carbon dioxide, in $cm^3\ min^{-1}$, between 2 and 8 minutes?

   A 5

   B 26

   C 30

   D 41

6. 5 g of copper is added to excess silver nitrate solution. The equation for the reaction that takes place is:

   $$Cu(s) + 2AgNO_3(aq) \rightarrow 2Ag(s) + Cu(NO_3)_2(aq)$$

   After some time, the solid present is filtered off from the solution, washed with water, dried and weighed.

   The final mass of the solid will be

   A less than 5 g

   B 5 g

   C 10 g

   D more than 10 g.

7.

   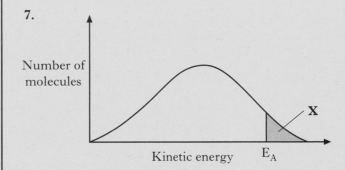

   Number of molecules

   Kinetic energy E_A

   In area **X**

   A molecules always form an activated complex

   B no molecules have the energy to form an activated complex

   C collisions between molecules are always successful in forming products

   D all molecules have the energy to form an activated complex.

8.

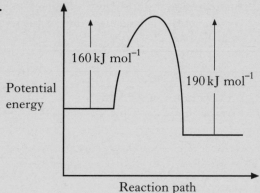

When a catalyst is used, the activation energy of the forward reaction is reduced to $35\,kJ\,mol^{-1}$.

What is the activation energy of the catalysed reverse reaction?

A $30\,kJ\,mol^{-1}$

B $35\,kJ\,mol^{-1}$

C $65\,kJ\,mol^{-1}$

D $190\,kJ\,mol^{-1}$

9. As the atomic number of the alkali metals increases

A   the first ionisation energy decreases

B   the atomic size decreases

C   the density decreases

D   the melting point increases.

10. Which of the following atoms has the least attraction for bonding electrons?

A   Carbon

B   Nitrogen

C   Phosphorus

D   Silicon

11. Which of the following reactions refers to the third ionisation energy of aluminium?

A $Al(s) \rightarrow Al^{3+}(g) + 3e^-$

B $Al(g) \rightarrow Al^{3+}(g) + 3e^-$

C $Al^{2+}(g) \rightarrow Al^{3+}(g) + e^-$

D $Al^{3+}(g) \rightarrow Al^{4+}(g) + e^-$

12. Which of the following represents an exothermic process?

A $Cl_2(g) \rightarrow 2Cl(g)$

B $Na(s) \rightarrow Na(g)$

C $Na(g) \rightarrow Na^+(g) + e^-$

D $Na^+(g) + Cl^-(g) \rightarrow Na^+Cl^-(s)$

13. In which of the following liquids does hydrogen bonding occur?

A   Ethanol

B   Ethyl ethanoate

C   Hexane

D   Pent-1-ene

14. The shapes of some common molecules are shown. Each molecule contains at least one polar covalent bond.

Which of the following molecules is non-polar?

A $H - Cl$

B

C $O = C = O$

D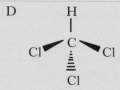

15. At room temperature, a solid substance was shown to have a lattice consisting of positively charged ions and delocalised outer electrons.

The substance could be

A   graphite

B   sodium

C   mercury

D   phosphorus.

16. The mass of 1 mol of sodium is 23 g.

What is the approximate mass of one sodium atom?

A 6×10^{23} g

B 6×10^{-23} g

C $3 \cdot 8 \times 10^{-23}$ g

D $3 \cdot 8 \times 10^{-24}$ g

17. In which of the following pairs do the gases contain the same number of oxygen atoms?

A 1 mol of oxygen and 1 mol of carbon monoxide

B 1 mol of oxygen and 0·5 mol of carbon dioxide

C 0·5 mol of oxygen and 1 mol of carbon dioxide

D 1 mol of oxygen and 1 mol of carbon dioxide

18. The Avogadro Constant is the same as the number of

A molecules in 16 g of oxygen

B electrons in 1 g of hydrogen

C atoms in 24 g of carbon

D ions in 1 litre of sodium chloride solution, concentration $1 \, \text{mol} \, l^{-1}$.

19. $$2NO(g) + O_2(g) \rightarrow 2NO_2(g)$$

How many litres of nitrogen dioxide gas would be produced in a reaction, starting with a mixture of 5 litres of nitrogen monoxide gas and 2 litres of oxygen gas?

(All volumes are measured under the same conditions of temperature and pressure.)

A 2

B 3

C 4

D 5

20. Which of the following fuels can be produced by the fermentation of biological material under anaerobic conditions?

A Hydrogen

B Methane

C Methanol

D Petrol

21. Butadiene is the first member of a homologous series of hydrocarbons called dienes.

What is the general formula for this series?

A C_nH_{n+2}

B C_nH_{n+3}

C C_nH_{2n}

D C_nH_{2n-2}

22.
$$CH_3 - CH = CH_2$$

Reaction **X** ↓

$$CH_3 - CH_2 - CH_2 - OH$$

Reaction **Y** ↓

$$CH_3 - CH_2 - C \underset{H}{\overset{O}{\lessgtr}}$$

Which line in the table correctly describes reactions **X** and **Y**?

	Reaction X	Reaction Y
A	hydration	oxidation
B	hydration	reduction
C	hydrolysis	oxidation
D	hydrolysis	reduction

23. Ammonia is manufactured from hydrogen and nitrogen by the Haber Process.

$$3H_2(g) \; + \; N_2(g) \; \rightleftharpoons \; 2NH_3(g)$$

If 80 kg of ammonia is produced from 60 kg of hydrogen, what is the percentage yield?

A $\dfrac{80}{340} \times 100$

B $\dfrac{80}{170} \times 100$

C $\dfrac{30}{80} \times 100$

D $\dfrac{60}{80} \times 100$

[Turn over

24. Which of the following statements about methanol is **false**?

A   It can be made from synthesis gas.

B   It can be dehydrated to form an alkene.

C   It can be oxidised to give a carboxylic acid.

D   It reacts with acidified potassium dichromate solution.

25. A by-product produced in the manufacture of a polyester has the structure shown.

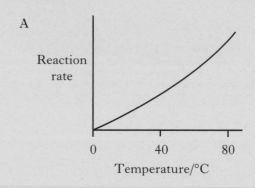

What is the structure of the diacid monomer used in the polymerisation?

A   COOH

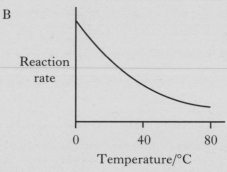

COOH

B   COOH

COOH

C $HOOC - CH_2 - CH_2 - COOH$

D   COOH

CH₂CH₂COOH

26. Which of the following statements can be applied to polymeric esters?

A   They are used for flavourings, perfumes and solvents.

B   They are manufactured for use as textile fibres and resins.

C   They are cross-linked addition polymers.

D   They are condensation polymers made by the linking up of amino acids.

27. The rate of hydrolysis of protein, using an enzyme, was studied at different temperatures.

Which of the following graphs would be obtained?

A

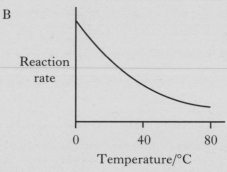

B

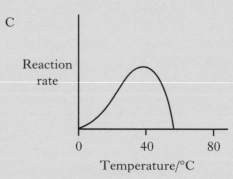

C

D

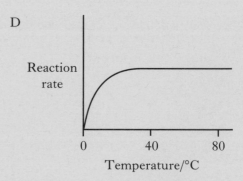

28. Which of the following arrangements of atoms shows a peptide link?

```
        H         H
        |         |
A   — C — O — N —
        |
        H


        H   O   H
        |   ||  |
B   — C — C — N —
        |
        H


        H   OH
        |   |
C   — C — C = N —
        |
        H


        H   O       H
        |   ||      |
D   — C — C — O — N —
        |
        H
```

29. Which line in the table shows the effect of a catalyst on the reaction rates and position of equilibrium in a reversible reaction?

	Rate of forward reaction	Rate of reverse reaction	Position of equilibrium
A	increased	unchanged	moves right
B	increased	increased	unchanged
C	increased	decreased	moves right
D	unchanged	unchanged	unchanged

30. The following equilibrium exists in bromine water.

$$Br_2(aq) + H_2O(\ell) \rightleftharpoons Br^-(aq) + 2H^+(aq) + OBr^-(aq)$$
(red)   (colourless)   (colourless)

The red colour of bromine water would fade on adding a few drops of a concentrated solution of

A   HCl

B   KBr

C $AgNO_3$

D   NaOBr.

31. Which of the following is the best description of a 0.1 mol l^{-1} solution of nitric acid?

A   Dilute solution of a weak acid

B   Dilute solution of a strong acid

C   Concentrated solution of a weak acid

D   Concentrated solution of a strong acid

32. The conductivity of pure water is low because

A   water molecules are polar

B   only a few water molecules are ionised

C   water molecules are linked by hydrogen bonds

D   there are equal numbers of hydrogen and hydroxide ions in water.

33. Which of the following statements is **true** about an aqueous solution of ammonia?

A   It has a pH less than 7.

B   It is completely ionised.

C   It contains more hydroxide ions than hydrogen ions.

D   It reacts with acids producing ammonia gas.

34. Equal volumes of solutions of ethanoic acid and hydrochloric acid, of equal concentration, are compared.

In which of the following cases does the ethanoic acid give the higher value?

A   pH of solution

B   Conductivity of solution

C   Rate of reaction with magnesium

D   Volume of sodium hydroxide solution neutralised

35. Equal volumes of 0.1 mol l^{-1} solutions of the following acids and alkalis were mixed.

Which of the following pairs would give the solution with the lowest pH?

A   Hydrochloric acid and sodium hydroxide

B   Hydrochloric acid and calcium hydroxide

C   Sulphuric acid and sodium hydroxide

D   Sulphuric acid and calcium hydroxide

[Turn over

36. Which of the following compounds dissolves in water to form an acidic solution?

    A Sodium nitrate

    B Barium sulphate

    C Potassium ethanoate

    D Ammonium chloride

37. The ion-electron equations for a redox reaction are:

$$2I^-(aq) \rightarrow I_2(aq) + 2e^-$$

$$MnO_4^-(aq) + 8H^+(aq) + 5e^- \rightarrow Mn^{2+}(aq) + 4H_2O(\ell)$$

How many moles of iodide ions are oxidised by one mole of permanganate ions?

    A 0·2

    B 0·4

    C 2

    D 5

38. In which of the following reactions is the hydrogen ion acting as an oxidising agent?

    A $Mg + 2HCl \rightarrow MgCl_2 + H_2$

    B $NaOH + HNO_3 \rightarrow NaNO_3 + H_2O$

    C $CuCO_3 + H_2SO_4 \rightarrow CuSO_4 + H_2O + CO_2$

    D $CH_3COONa + HCl \rightarrow NaCl + CH_3COOH$

39. An atom of ^{227}Th decays by a series of alpha emissions to form an atom of ^{211}Pb.

How many alpha particles are released in the process?

    A 2

    B 3

    C 4

    D 5

40. The half-life of the isotope ^{210}Pb is 21 years.

What fraction of the original ^{210}Pb atoms will be present after 63 years?

    A 0·5

    B 0·25

    C 0·125

    D 0·0625

Candidates are reminded that the answer sheet MUST be returned INSIDE the front cover of this answer book.

DO NOT WRITE IN THIS MARGIN

Marks

SECTION B

All answers must be written clearly and legibly in ink.

1. The formulae for three oxides of sodium, carbon and silicon are Na_2O, CO_2 and SiO_2.

Complete the table for CO_2 and SiO_2 to show both the bonding and structure of the three oxides at room temperature.

Oxide	Bonding and structure
Na_2O	ionic lattice
CO_2	
SiO_2	

(2)

2. A typical triglyceride found in olive oil is shown below.

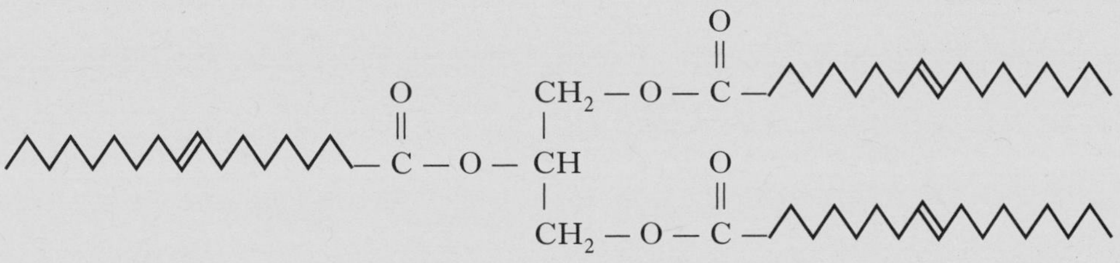

(*a*) To which family of organic compounds do triglycerides belong? 1

(*b*) Olive oil can be hardened for use in margarines.

What happens to the triglyceride molecules during the hardening of olive oil? 1

(*c*) Give **one** reason why oils can be a useful part of a balanced diet. 1

(3)

Marks

3. A student carried out the Prescribed Practical Activity (PPA) to find the effect of concentration on the rate of the reaction between hydrogen peroxide solution and an acidified solution of iodide ions.

$$H_2O_2(aq) \quad + \quad 2H^+(aq) \quad + \quad 2I^-(aq) \quad \rightarrow \quad 2H_2O(\ell) \quad + \quad I_2(aq)$$

During the investigation, only the concentration of the iodide ions was changed.

Part of the student's results sheet for this PPA is shown.

Results

Experiment	Volume of KI(aq) /cm³	Volume of H₂O /cm³	Volume of H₂O₂(aq) /cm³	Volume of H₂SO₄(aq) /cm³	Volume of Na₂S₂O₃(aq) /cm³	Rate /s⁻¹
1	25	0	5	10	10	0·043
2						
3						

(a) Describe how the concentration of the potassium iodide solution was changed during this series of experiments.

1

(b) Calculate the reaction time, in seconds, for the first experiment.

1

(2)

Marks

4. Using a cobalt catalyst, alkenes react with a mixture of hydrogen and carbon monoxide.

The products are two isomeric aldehydes.

Propene reacts with the mixture as shown.

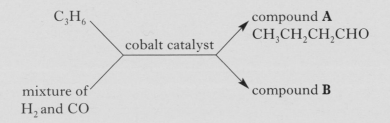

(a) What name is given to a mixture of hydrogen and carbon monoxide?

        1

(b) Draw a structural formula for compound **B**.

        1

(c) (i) What would be observed if compound **A** was gently heated with Tollens' reagent?

        1

    (ii) How would the reaction mixture be heated?

        1

(d) Aldehydes can also be formed by the reaction of some alcohols with copper(II) oxide.

Name the **type** of alcohol that would react with copper(II) oxide to form an aldehyde.

        1

        (5)

[Turn over

Marks

5. All the isotopes of technetium are radioactive.

(*a*) Technetium-99 is produced as shown.

$$^{99}_{42}\text{Mo} \rightarrow {}^{99}_{43}\text{Tc} + \textbf{X}$$

Identify **X**.

1

(*b*) The graph shows the decay curve for a 1·0 g sample of technetium-99.

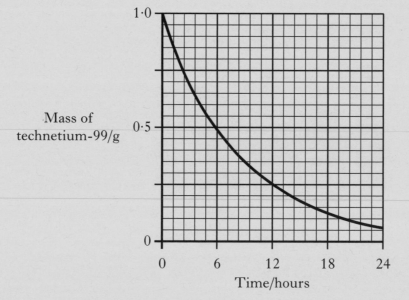

Mass of technetium-99/g

Time/hours

(i) Draw a curve on the graph to show the variation of mass with time for a 0·5 g sample of technetium-99.

(An additional graph, if required, can be found on *Page twenty-eight*.)

1

(ii) Technetium-99 is widely used in medicine to detect damage to heart tissue. It is a gamma-emitting radioisotope and is injected into the body.

Suggest **one** reason why technetium-99 can be safely used in this way.

1

(3)

DO NOT WRITE IN THIS MARGIN

Marks

6. In 1865, the German chemist Kekulé proposed a ring structure for benzene.

This structure was based on alternating single and double bonds.

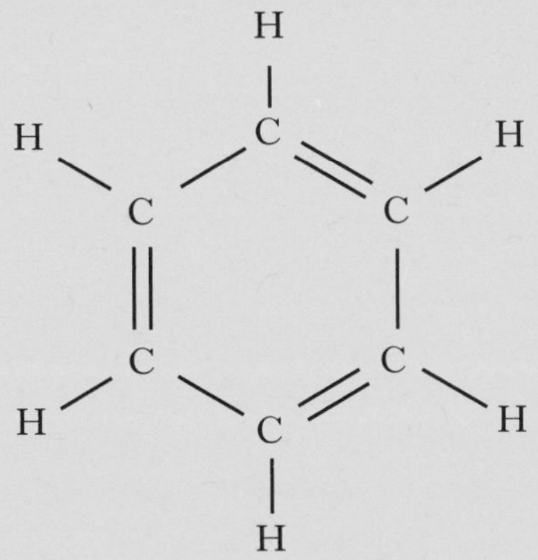

(*a*) (i) Describe a chemical test that would indicate that the above chemical structure for benzene is incorrect. **1**

(ii) Briefly describe the correct structure for benzene. **1**

(*b*) Benzene can be formed from cyclohexane.

$$C_6H_{12} \rightarrow C_6H_6 + 3H_2$$

What name is given to this type of reaction? **1**

(*c*) Benzene is also added in very small amounts to some petrols.

Why is benzene added to petrol? **1**

(4)

Marks

7. Hydrogen fluoride, HF, is used to manufacture hydrofluorocarbons.

Hydrofluorocarbons are now used as refrigerants instead of chlorofluorocarbons, CFCs.

(*a*) Why are CFCs no longer used?

1

(*b*) Hydrogen fluoride gas is manufactured by reacting calcium fluoride with concentrated sulphuric acid.

$$CaF_2 \ + \ H_2SO_4 \ \rightarrow \ CaSO_4 \ + \ 2HF$$

What volume of hydrogen fluoride gas is produced when $1.0\,kg$ of calcium fluoride reacts completely with concentrated sulphuric acid?

(Take the molar volume of hydrogen fluoride gas to be 24 litres mol^{-1}.)

Show your working clearly.

2

(3)

Marks

8. Carbon monoxide can be produced in many ways.

(*a*) One method involves the reaction of carbon with an oxide of boron.

$$B_2O_3 \quad + \quad C \quad \rightarrow \quad B_4C \quad + \quad CO$$

Balance this equation.

1

(*b*) Carbon monoxide is also a product of the reaction of carbon dioxide with hot carbon. The carbon dioxide is made by the reaction of dilute hydrochloric acid with solid calcium carbonate.

Unreacted carbon dioxide is removed before the carbon monoxide is collected by displacement of water.

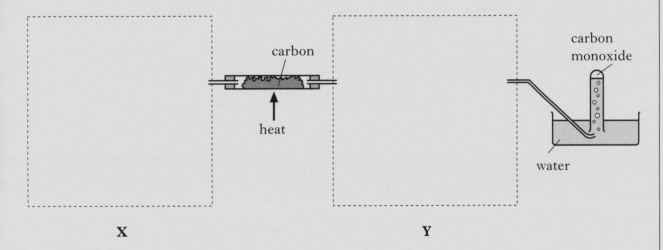

X   **Y**

Complete the diagram to show how the carbon dioxide can be produced at **X** and how the unreacted carbon dioxide can be removed by bubbling it through a solution at **Y**.

Normal laboratory apparatus should be used in your answer and the chemicals used at **X** and **Y** should be labelled.

2

(*c*) Why is carbon monoxide present in car exhaust fumes?

1

(4)

[Turn over

Marks

9. Hydrogen gas has a boiling point of −253 °C.

(a) Explain clearly why hydrogen is a gas at room temperature.

In your answer you should name the intermolecular forces involved and indicate how they arise.

2

(b) Hydrogen gas can be prepared in the lab by the electrolysis of dilute sulphuric acid.

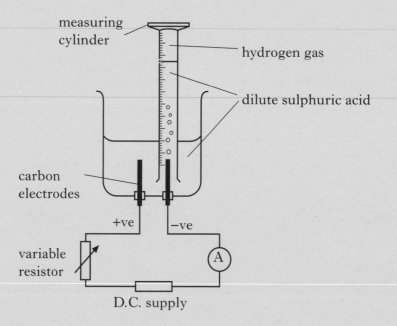

(i) Before collecting the gas in the measuring cylinder, it is usual to switch on the current and allow bubbles of gas to be produced for a few minutes.

Why is this done?

1

Marks

9. (*b*) **(continued)**

(ii) The equation for the reaction at the negative electrode is:

$$2H^+(aq) \quad + \quad 2e^- \quad \rightarrow \quad H_2(g)$$

Calculate the mass of hydrogen gas, in grams, produced in 10 minutes when a current of $0 \cdot 30$ A was used.

Show your working clearly.

2

(*c*) The concentration of $H^+(aq)$ ions in the dilute sulphuric acid used in the experiment was 1×10^{-1} mol l^{-1}.

Calculate the concentration of $OH^-(aq)$ ions, in mol l^{-1}, in the dilute sulphuric acid.

1

(6)

[Turn over

Marks

10. When cyclopropane gas is heated over a catalyst, it isomerises to form propene gas and an equilibrium is obtained.

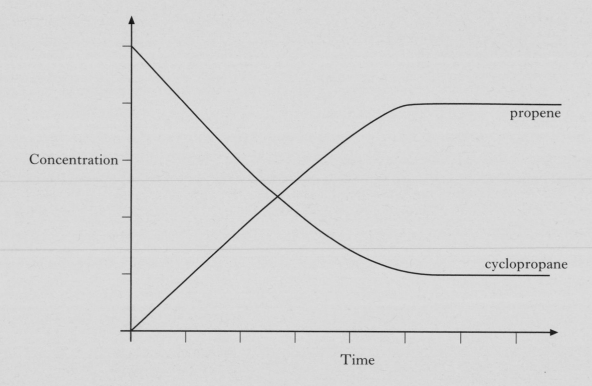

cyclopropane   propene

The graph shows the concentrations of cyclopropane and propene as equilibrium is established in the reaction.

(*a*) Mark clearly on the graph the point at which equilibrium has just been reached.

1

(*b*) Why does increasing the pressure have **no** effect on the position of this equilibrium?

1

Marks

10. (continued)

(*c*) The equilibrium can also be achieved by starting with propene.

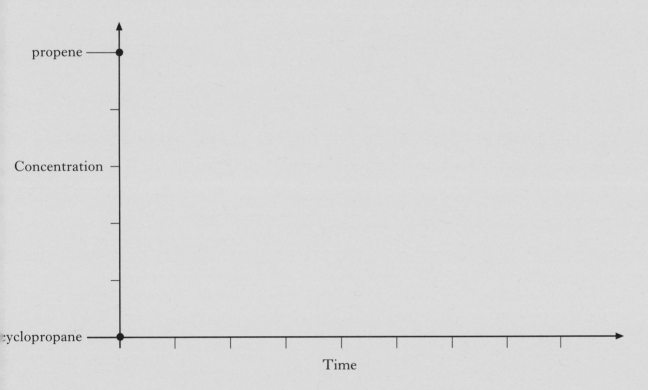

Using the initial concentrations shown, sketch a graph to show how the concentrations of propene and cyclopropane change as equilibrium is reached for this reverse reaction.

1

(3)

[Turn over

Marks

11. A student writes the following two statements. **Both are incorrect**.
In each case explain the mistake in the student's reasoning.

(a) Alcohols are alkaline because of their OH groups.

1

(b) Because of the iodine, potassium iodide will produce a blue/black colour in contact with starch.

1

(2)

Marks

12. When in danger, bombardier beetles can fire a hot, toxic mixture of chemicals at the attacker.

This mixture contains quinone, $C_6H_4O_2$, a compound that is formed by the reaction of hydroquinone, $C_6H_4(OH)_2$, with hydrogen peroxide, H_2O_2. The reaction is catalysed by an enzyme called catalase.

(a) Most enzymes can catalyse only specific reactions, eg catalase cannot catalyse the hydrolysis of starch.

Give a reason for this.

1

(b) The equation for the overall reaction is:

$$C_6H_4(OH)_2(aq) \;+\; H_2O_2(aq) \;\rightarrow\; C_6H_4O_2(aq) \;+\; 2H_2O(\ell)$$

Use the following data to calculate the enthalpy change, in $kJ\,mol^{-1}$, for the above reaction.

$$C_6H_4(OH)_2(aq) \;\rightarrow\; C_6H_4O_2(aq) \;+\; H_2(g) \qquad \Delta H = +177{\cdot}4\,kJ\,mol^{-1}$$

$$H_2(g) \;+\; O_2(g) \;\rightarrow\; H_2O_2(aq) \qquad \Delta H = -191{\cdot}2\,kJ\,mol^{-1}$$

$$H_2(g) \;+\; \tfrac{1}{2}O_2(g) \;\rightarrow\; H_2O(g) \qquad \Delta H = -241{\cdot}8\,kJ\,mol^{-1}$$

$$H_2O(g) \;\rightarrow\; H_2O(\ell) \qquad \Delta H = -43{\cdot}8\,kJ\,mol^{-1}$$

Show your working clearly.

2

(3)

[Turn over

Marks

13. For many years, carbohydrates found in plants have been used to provide chemicals. Lactic acid can be produced by fermenting the carbohydrates in corn.

Lactic acid has the structure:

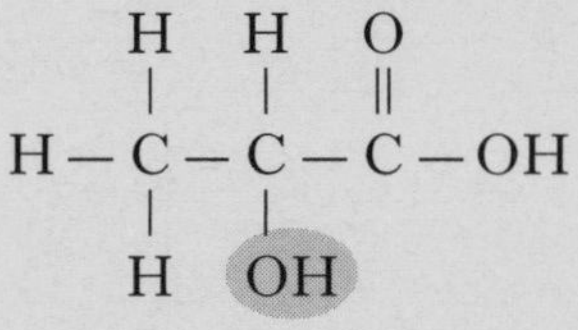

$$\begin{array}{ccc} H & H & O \\ | & | & \| \\ H-C-C-C-OH \\ | & | \\ H & OH \end{array}$$

(a) Name the functional group in the shaded area.

1

(b) Lactic acid is used to make polylactic acid, a biodegradeable polymer that is widely used for food packaging.

(i) Name another biodegradeable polymer.

1

(ii) Polylactic acid can be manufactured by either a batch or a continuous process.

What is meant by a batch process?

1

(iii) The first stage in the polymerisation of lactic acid involves the condensation of two lactic acid molecules to form a cyclic structure called a lactone.

Draw a structural formula for the lactone formed when two molecules of lactic acid undergo condensation with each other.

1

Marks

14. Hydrogen peroxide decomposes as shown:

$$H_2O_2(aq) \rightarrow H_2O(\ell) + \tfrac{1}{2}O_2(g)$$

The reaction can be catalysed by iron(III) nitrate solution.

(a) What **type** of catalyst is iron(III) nitrate solution in this reaction?

1

(b) In order to calculate the enthalpy change for the decomposition of hydrogen peroxide, a student added iron(III) nitrate solution to hydrogen peroxide solution.

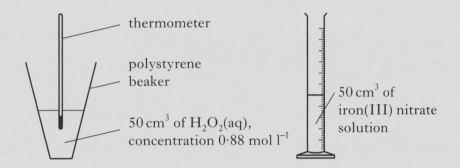

thermometer

polystyrene beaker

$50\,cm^3$ of $H_2O_2(aq)$, concentration $0{\cdot}88\,mol\,l^{-1}$

$50\,cm^3$ of iron(III) nitrate solution

As a result of the reaction, the temperature of the solution in the polystyrene beaker increased by 16 °C.

(i) What is the effect of the catalyst on the enthalpy change (ΔH) for the reaction?

1

(ii) Use the experimental data to calculate the enthalpy change, in $kJ\,mol^{-1}$, for the decomposition of hydrogen peroxide.
Show your working clearly.

3

(5)

DO NOT
WRITE IN
THIS
MARGIN

Marks

15. (*a*) The graph shows how the freezing point changes with changing concentration for aqueous solutions of sodium chloride and ethane-1,2-diol.

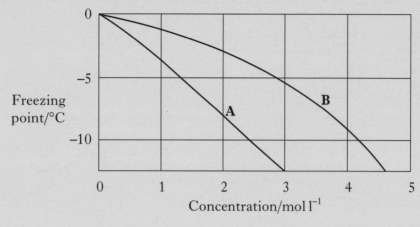

A sodium chloride

B ethane-1,2-diol

(i) Draw a structural formula for ethane-1,2-diol.

1

(ii) Ethane-1,2-diol solution is used as an antifreeze in car radiators, yet from the graph it would appear that sodium chloride solution is more efficient.

Suggest why sodium chloride solution is **not** used as an antifreeze.

1

(*b*) Boiling points can be used to compare the strengths of the intermolecular forces in alkanes with the strengths of the intermolecular forces in diols.

Name the alkane that should be used to make a valid comparison between the strength of its intermolecular forces and those in ethane-1,2-diol.

1

(3)

Marks

16. Aldehydes and ketones can take part in a reaction sometimes known as an aldol condensation.

The simplest aldol reaction involves two molecules of ethanal.

$$
\begin{array}{c}
\quad H \quad H \\
\quad | \quad\; | \\
H-C-C=O \\
\quad | \\
\quad H
\end{array}
\;+\;
\begin{array}{c}
\quad H \quad H \\
\quad | \quad\; | \\
H-C-C=O \\
\quad | \\
\quad H
\end{array}
\longrightarrow
\begin{array}{c}
\quad H \quad H \quad H \quad H \\
\quad | \quad\; | \quad\; | \quad\; | \\
H-C-C-C-C=O \\
\quad | \quad\; | \quad\; | \\
\quad H \quad OH \quad H
\end{array}
$$

In the reaction, the carbon atom next to the carbonyl functional group of one molecule forms a bond with the carbonyl carbon atom of the second molecule.

(*a*) Draw a structural formula for the product formed when propanone is used instead of ethanal in this type of reaction.

1

(*b*) Name an aldehyde that would **not** take part in an aldol condensation.

1

(*c*) Apart from the structure of the reactants, suggest what is unusual about applying the term "condensation" to this particular type of reaction.

1

(3)

[Turn over

Marks

17. Oxalic acid is found in rhubarb. The number of moles of oxalic acid in a carton of rhubarb juice can be found by titrating samples of the juice with a solution of potassium permanganate, a powerful oxidising agent.

The equation for the overall reaction is:

$$5(COOH)_2(aq) + 6H^+(aq) + 2MnO_4^-(aq) \rightarrow 2Mn^{2+}(aq) + 10CO_2(aq) + 8H_2O(\ell)$$

(a) Write the ion-electron equation for the reduction reaction.

1

(b) Why is an indicator **not** required to detect the end-point of the titration?

1

(c) In an investigation using a $500\,cm^3$ carton of rhubarb juice, separate $25.0\,cm^3$ samples were measured out. Three samples were then titrated with $0.040\,mol\,l^{-1}$ potassium permanganate solution, giving the following results.

Titration	Volume of potassium permanganate solution used/cm^3
1	27·7
2	26·8
3	27·0

Average volume of potassium permanganate solution used = $26.9\,cm^3$.

(i) Why was the first titration result not included in calculating the average volume of potassium permanganate solution used?

1

Marks

17. (*c*) **(continued)**

(ii) Calculate the number of moles of oxalic acid in the $500 \, cm^3$ carton of rhubarb juice.

Show your working clearly.

2

(5)

[*END OF QUESTION PAPER*]

SPACE FOR ANSWERS

ADDITIONAL GRAPH FOR QUESTION 5(b)(i)

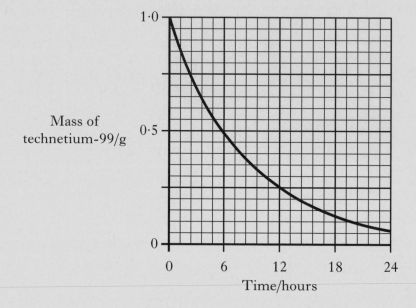

ADDITIONAL SPACE FOR ANSWERS

ADDITIONAL SPACE FOR ANSWERS

[BLANK PAGE]

FOR OFFICIAL USE

Total
Section B

X012/301

NATIONAL
QUALIFICATIONS
2009

WEDNESDAY, 3 JUNE
9.00 AM – 11.30 AM

CHEMISTRY
HIGHER

Fill in these boxes and read what is printed below.

Full name of centre

Town

Forename(s)

Surname

Date of birth

Day Month Year   Scottish candidate number   Number of seat

Reference may be made to the Chemistry Higher and Advanced Higher Data Booklet.

SECTION A—Questions 1–40 (40 marks)

Instructions for completion of **Section A** are given on page two.

For this section of the examination you must use an **HB pencil**.

SECTION B (60 marks)

1   All questions should be attempted.

2   The questions may be answered in any order but all answers are to be written in the spaces provided in this answer book, **and must be written clearly and legibly in ink**.

3   Rough work, if any should be necessary, should be written in this book and then scored through when the fair copy has been written. If further space is required, a supplementary sheet for rough work may be obtained from the invigilator.

4   Additional space for answers will be found at the end of the book. If further space is required, supplementary sheets may be obtained from the invigilator and should be inserted inside the **front** cover of this book.

5   The size of the space provided for an answer should not be taken as an indication of how much to write. It is not necessary to use all the space.

6   Before leaving the examination room you must give this book to the invigilator. If you do not, you may lose all the marks for this paper.

SECTION A

Read carefully

1 Check that the answer sheet provided is for **Chemistry Higher (Section A)**.

2 For this section of the examination you must use an **HB pencil** and, where necessary, an eraser.

3 Check that the answer sheet you have been given has **your name**, **date of birth**, **SCN** (Scottish Candidate Number) and **Centre Name** printed on it.

Do not change any of these details.

4 If any of this information is wrong, tell the Invigilator immediately.

5 If this information is correct, **print** your name and seat number in the boxes provided.

6 The answer to each question is **either** A, B, C or D. Decide what your answer is, then, using your pencil, put a horizontal line in the space provided (see sample question below).

7 There is **only one correct** answer to each question.

8 Any rough working should be done on the question paper or the rough working sheet, **not** on your answer sheet.

9 At the end of the exam, put the **answer sheet for Section A inside the front cover of your answer book**.

Sample Question

To show that the ink in a ball-pen consists of a mixture of dyes, the method of separation would be

        A chromatography

        B fractional distillation

        C fractional crystallisation

        D filtration.

The correct answer is **A**—chromatography. The answer **A** has been clearly marked in **pencil** with a horizontal line (see below).

Changing an answer

If you decide to change your answer, carefully erase your first answer and using your pencil, fill in the answer you want. The answer below has been changed to **D**.

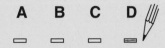

1. Which of the following oxides forms an aqueous solution with pH greater than 7?

   A   Carbon dioxide

   B   Copper(II) oxide

   C   Sulphur dioxide

   D   Sodium oxide

2. In which of the following reactions is a positive ion reduced?

   A   Iodide $\longrightarrow$ iodine

   B   Nickel(II) $\longrightarrow$ nickel(III)

   C   Cobalt(III) $\longrightarrow$ cobalt(II)

   D   Sulphate $\longrightarrow$ sulphite

3. Which of the following elements is most likely to have a covalent network structure?

Element	Melting point/°C	Boiling point/°C	Density/ $g\,cm^{-3}$	Conduction when solid
A	44	280	1·82	No
B	660	2467	2·70	Yes
C	1410	2355	2·33	No
D	114	184	4·93	No

4. Two identical samples of copper(II) carbonate were added to an excess of $1\ mol\,l^{-1}$ hydrochloric acid and $1\ mol\,l^{-1}$ sulphuric acid respectively.

   Which of the following would have been different for the two reactions?

   A   The pH of the final solution

   B   The volume of gas produced

   C   The mass of water formed

   D   The mass of copper(II) carbonate dissolved

5. The graph shows how the rate of a reaction varies with the concentration of one of the reactants.

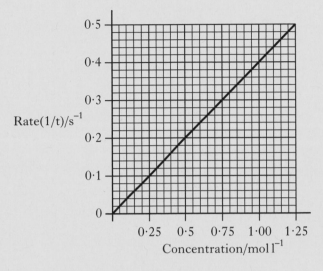

   What was the reaction time, in seconds, when the concentration of the reactant was $0·50\ mol\,l^{-1}$?

   A   0·2

   B   0·5

   C   2·0

   D   5·0

6. 10 g of magnesium is added to 1 litre of $1\ mol\,l^{-1}$ copper(II) sulphate solution and the mixture stirred until no further reaction occurs.

   Which of the following is a result of this reaction?

   A   All the magnesium reacts.

   B   63·5 g of copper is displaced.

   C   2 mol of copper is displaced.

   D   The resulting solution is colourless.

   [Turn over

7. A reaction was carried out with and without a catalyst as shown in the energy diagram.

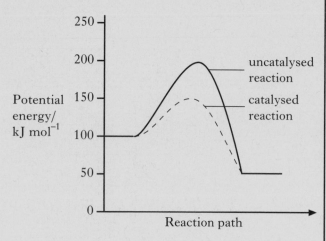

What is the enthalpy change, in $kJ\,mol^{-1}$, for the catalysed reaction?

A   −100

B   −50

C   +50

D   +100

8. Ethanol (C_2H_5OH) has a different enthalpy of combustion from dimethyl ether (CH_3OCH_3). This is because the compounds have different

A   boiling points

B   molecular masses

C   products of combustion

D   bonds within the molecules.

9. Which of the following compounds has the greatest ionic character?

A   Caesium fluoride

B   Caesium iodide

C   Sodium fluoride

D   Sodium iodide

10. Which line in the table is likely to be correct for the element francium?

	State at 30 °C	First ionisation energy/kJ mol^{-1}
A	solid	less than 382
B	liquid	less than 382
C	solid	greater than 382
D	liquid	greater than 382

11. Which of the following equations represents the first ionisation energy of fluorine?

A $F^-(g) \rightarrow F(g) + e^-$

B $F^-(g) \rightarrow \frac{1}{2}F_2(g) + e^-$

C $F(g) \rightarrow F^+(g) + e^-$

D $\frac{1}{2}F_2(g) \rightarrow F^+(g) + e^-$

12. The two hydrogen atoms in a molecule of hydrogen are held together by

A   a hydrogen bond

B   a polar covalent bond

C   a non-polar covalent bond

D   a van der Waals' force.

13. In which of the following compounds would hydrogen bonding **not** occur?

A
$$\begin{array}{ccc} H & H & H \\ | & | & | \\ H-C-C-N-H \\ | & | \\ H & H \end{array}$$

B
$$\begin{array}{cc} H & H \\ | & | \\ H-C-C-O-H \\ | & | \\ H & H \end{array}$$

C
$$\begin{array}{ccc} H & H & H \\ | & | & | \\ H-C-N-C-H \\ | & | \\ H & H \end{array}$$

D
$$\begin{array}{cc} H & H \\ | & | \\ H-C-O-C-H \\ | & | \\ H & H \end{array}$$

14. Which of the following shows the types of bonding in **decreasing** order of strength?

A   Covalent : hydrogen : van der Waals'

B   Covalent : van der Waals' : hydrogen

C   Hydrogen : covalent : van der Waals'

D   Van der Waals' : hydrogen : covalent

15. What type of bonding and structure is found in a fullerene?

   A   Ionic lattice

   B   Metallic lattice

   C   Covalent network

   D   Covalent molecular

16. Some covalent compounds are made up of molecules that contain polar bonds but the molecules are overall non-polar.

   Which of the following covalent compounds is made up of non-polar molecules?

   A   Ammonia

   B   Water

   C   Carbon tetrachloride

   D   Hydrogen fluoride

17. The Avogadro Constant is the same as the number of

   A   ions in 1 mol of NaCl

   B   atoms in 1 mol of hydrogen gas

   C   electrons in 1 mol of helium gas

   D   molecules in 1 mol of oxygen gas.

18. Which of the following gas samples has the same volume as 7 g of carbon monoxide?

   (All volumes are measured at the same temperature and pressure.)

   A   1 g of hydrogen

   B   3·5 g of nitrogen

   C   10 g of argon

   D   35·5 g of chlorine

19. What volume of oxygen (in litres) would be required for the complete combustion of a gaseous mixture containing 1 litre of carbon monoxide and 3 litres of hydrogen?

   (All volumes are measured at the same temperature and pressure.)

   A   1

   B   2

   C   3

   D   4

20. Which of the following pollutants, produced during internal combustion in a car engine, is **not** the result of incomplete combustion?

   A   Carbon

   B   Carbon monoxide

   C   Hydrocarbons

   D   Nitrogen dioxide

21. Which of the following compounds does **not** have isomeric structures?

   A C_2HCl_3

   B $C_2H_4Cl_2$

   C   Propene

   D   Propan-1-ol

22. Which of the following compounds is an alkanone?

   A $CH_3 - CH_2 - \overset{\overset{\displaystyle O}{\|}}{C} - H$

   B $CH_3 - \overset{\overset{\displaystyle O}{\|}}{C} - O - CH_3$

   C $CH_3 - \overset{\overset{\displaystyle O}{\|}}{C} - CH_3$

   D $CH_3 - \overset{\overset{\displaystyle O}{\|}}{C} - OH$

23. What organic compound is produced by the dehydration of ethanol?

   A   Ethane

   B   Ethene

   C   Ethanal

   D   Ethanoic acid

24. The production of synthesis gas from methane involves

   A   steam reforming

   B   catalytic cracking

   C   hydration

   D   oxidation.

25. Compound **X** reacted with hot copper(II) oxide and the organic product did not give a colour change when heated with Fehling's solution.

Compound **X** could be

A butan-1-ol

B butan-2-ol

C butanone

D butanoic acid.

26. Part of a polymer is shown.

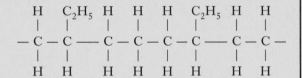

Which two alkenes were used to make this polymer?

A Ethene and propene

B Ethene and but-1-ene

C Propene and but-1-ene

D Ethene and but-2-ene

27. Ammonia solution may be used to distinguish $Fe^{2+}(aq)$ from $Fe^{3+}(aq)$ as follows:

$Fe^{2+}(aq)$ gives a green precipitate of $Fe(OH)_2$;

$Fe^{3+}(aq)$ gives a brown precipitate of $Fe(OH)_3$.

Which of the following types of compound is most likely to give similar results if used instead of ammonia?

A An alcohol

B An aldehyde

C An amine

D A carboxylic acid

28. Which of the following reactions takes place during the 'hardening' of vegetable oil?

A Addition

B Hydrolysis

C Dehydration

D Oxidation

29. Fats are formed by the condensation reaction between glycerol molecules and fatty acid molecules.

In this reaction the mole ratio of glycerol molecules to fatty acid molecules is

A 1 : 1

B 1 : 2

C 1 : 3

D 3 : 1.

30. Which of the following graphs shows how the rate of reaction varies with temperature for the fermentation of glucose?

A
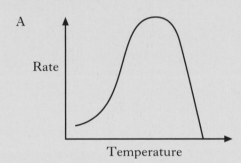

B
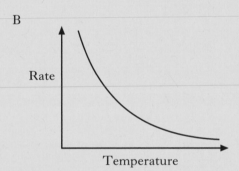

C
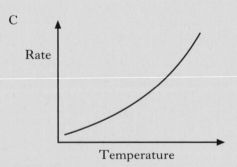

D
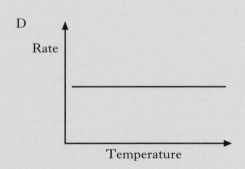

31. Which of the following is the best description of a feedstock?

A   A consumer product such as a textile, plastic or detergent.

B   A complex chemical that has been synthesised from small molecules.

C   A mixture of chemicals formed by the cracking of the naphtha fraction from oil.

D   A chemical from which other chemicals can be extracted or synthesised.

32.
$$S(s) + H_2(g) \rightarrow H_2S(g)$$
$$\Delta H = a$$

$$H_2(g) + \tfrac{1}{2}O_2(g) \rightarrow H_2O(\ell)$$
$$\Delta H = b$$

$$S(s) + O_2(g) \rightarrow SO_2(g)$$
$$\Delta H = c$$

$$H_2S(g) + 1\tfrac{1}{2}O_2(g) \rightarrow H_2O(\ell) + SO_2(g)$$
$$\Delta H = d$$

What is the relationship between a, b, c and d?

A $a = b + c - d$

B $a = d - b - c$

C $a = b - c - d$

D $a = d + c - b$

33. A catalyst is added to a reaction at equilibrium.

Which of the following does **not** apply?

A   The rate of the forward reaction increases.

B   The rate of the reverse reaction increases.

C   The position of equilibrium remains unchanged.

D   The position of equilibrium shifts to the right.

34. Steam and carbon monoxide react to form an equilibrium mixture.

$$CO(g) + H_2O(g) \rightleftharpoons H_2(g) + CO_2(g)$$

Which of the following graphs shows how the rates of the forward and reverse reactions change when carbon monoxide and steam are mixed?

KEY

——— forward reaction

············ reverse reaction

A

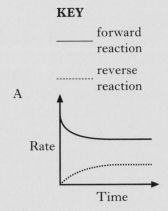

B

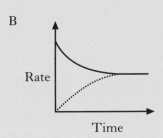

C

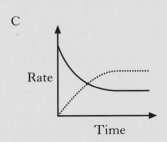

D

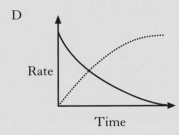

[Turn over

35. Solid sodium sulphite is dissolved in distilled water, producing an alkaline solution.

Which of the following processes is the most important in causing this change?

A Sodium ions reacting with hydroxide ions

B Hydrogen ions reacting with sulphite ions

C Sodium ions reacting with sulphite ions

D Hydrogen ions reacting with hydroxide ions

36. Which of the following salts dissolves in water to form an acidic solution?

A CH_3COONa

B Na_2SO_4

C KCl

D NH_4NO_3

37. Iodide ions can be oxidised using acidified potassium permanganate solution.

The equations are:

$$2I^-(aq) \rightarrow I_2(aq) + 2e^-$$

$$MnO_4^-(aq) + 8H^+(aq) + 5e^- \rightarrow Mn^{2+}(aq) + 4H_2O(\ell)$$

How many moles of iodide ions are oxidised by one mole of permanganate ions?

A 1·0

B 2·0

C 2·5

D 5·0

38. In the electrolysis of molten magnesium chloride, 1 mol of magnesium is deposited at the negative electrode by

A 96 500 coulombs

B 193 000 coulombs

C 1 mol of electrons

D 24·3 mol of electrons.

39. Alpha, beta and gamma radiation is passed from a source through an electric field onto a photographic plate.

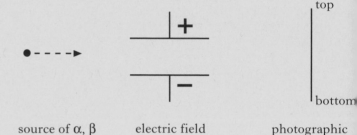

source of α, β   electric field   photographic
and γ radiations   plate

Which of the following patterns will be produced on the photographic plate?

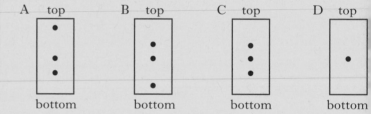

40. From which of the following could $^{32}_{15}P$ be produced by neutron capture?

A $^{33}_{15}P$

B $^{32}_{16}S$

C $^{31}_{15}P$

D $^{31}_{16}S$

**Candidates are reminded that the answer sheet MUST be returned INSIDE
the front cover of this answer book.**

[Turn over for Section B on *Page ten*

Marks

SECTION B

All answers must be written clearly and legibly in ink.

1. (a) Lithium starts the second period of the Periodic Table.

| Li | Be | B | C | N | O | F |

What is the trend in electronegativity values across this period from Li to F?

1

(b) **Graph 1** shows the first four ionisation energies for aluminium.

Graph 1

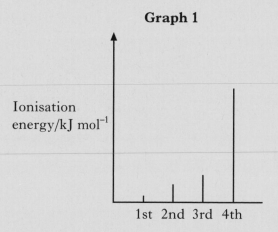

Ionisation
energy/kJ mol^{-1}

1st 2nd 3rd 4th

Why is the fourth ionisation energy of aluminium so much higher than the third ionisation energy?

1

Marks

1. **(continued)**

(c) **Graph 2** shows the boiling points of the elements in Group 7 of the Periodic Table.

Graph 2

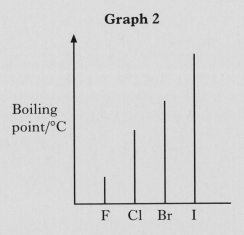

Why do the boiling points increase down Group 7?

1

(3)

[Turn over

DO NOT WRITE IN THIS MARGIN

Marks

2. Reactions are carried out in oil refineries to increase the octane number of petrol components. One such reaction is:

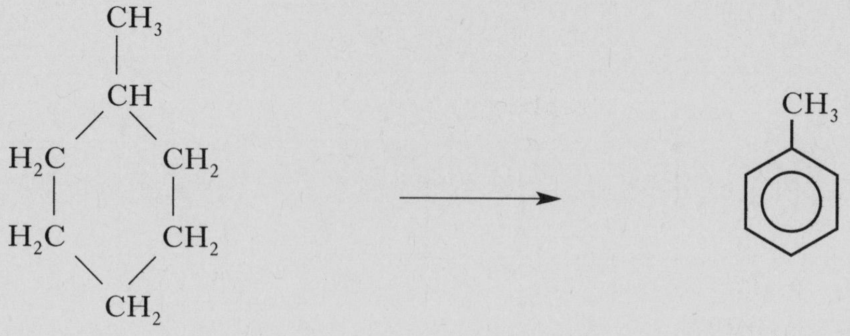

methylcyclohexane methylbenzene

(*a*) The molecular formula for methylbenzene can be written $C_{\mathbf{x}}H_{\mathbf{y}}$.

Give the values for **x** and **y**.

x =

y = **1**

(*b*) Hydrogen, the by-product of the reaction, can also be used as a fuel.

Give **one** advantage of using hydrogen as a fuel instead of petrol. **1**

(*c*) In some countries oxygenates are added to the petrol. These improve the efficiency of burning. One common oxygenate is ethanol.

Give **one** other advantage of adding ethanol to petrol. **1**

(3)

Marks

3. Alkanols can be oxidised to alkanoic acids.

$$CH_3CH_2CH_2OH \xrightarrow{\textbf{Step 1}} CH_3CH_2CHO \xrightarrow{\textbf{Step 2}} CH_3CH_2COOH$$

propan-1-ol   propanal   propanoic acid

(*a*)   (i)   Why can **Step 1** be described as an oxidation reaction?

1

     (ii)   Acidified potassium dichromate solution can be used to oxidise propanal in **Step 2**.

What colour change would be observed in this reaction?

1

(*b*)   Propan-1-ol and propanoic acid react to form an ester.

The mixture of excess reactants and ester product is poured onto sodium hydrogencarbonate solution.

  (i)   What evidence would show that an ester is formed?

1

   (ii)   Draw a structural formula for this ester.

1

(4)

[Turn over

4. Ozone gas, $O_3(g)$, is made up of triatomic molecules.

Marks

(a) Ozone is present in the upper atmosphere.

Why is the ozone layer important for life on Earth?

1

(b) The depletion of this layer is believed to be caused by chlorine radicals produced by the breakdown of certain CFCs. Some of the gaseous reactions are catalysed by ice crystals in clouds.

The crystals are acting as what type of catalyst?

1

(c) Ozone can be produced in the laboratory by electrical discharge.

$$3O_2(g) \rightarrow 2O_3(g)$$

Calculate the approximate number of $O_3(g)$ molecules produced from one mole of $O_2(g)$ molecules.

1

(3)

Marks

5. Polymers can be classified as natural or synthetic.

(a) Keratin, a natural polymer, is a protein found in hair.

The hydrolysis of keratin produces different monomers of the type shown.

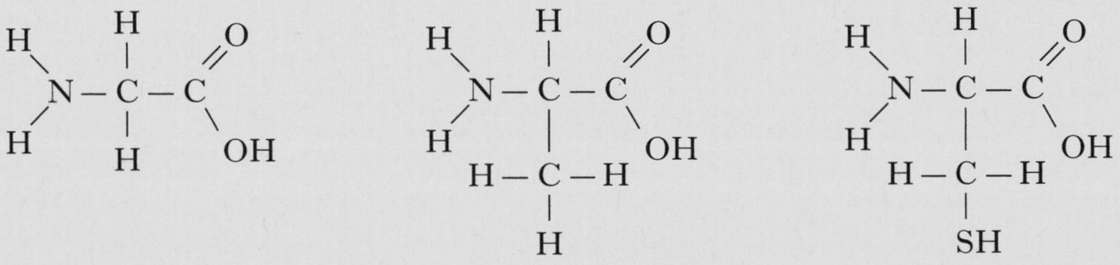

glycine   alanine   cysteine

(i) What name is given to monomers like glycine, alanine and cysteine?

1

(ii) What is meant by a **hydrolysis** reaction?

1

(b) Dacron, a synthetic polymer, is used in heart surgery.

A section of the polymer is shown.

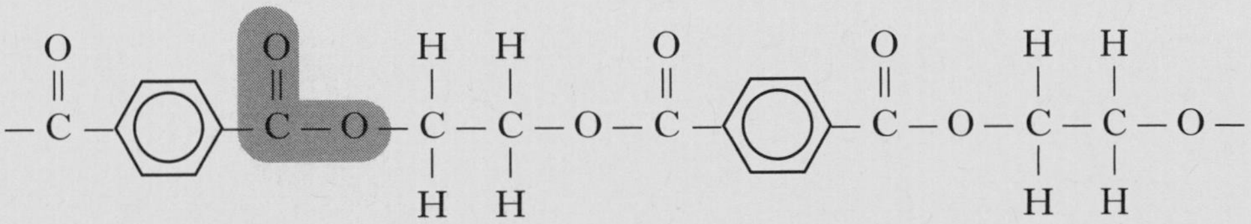

(i) What name is given to the link made by the shaded group of atoms in this section of the polymer?

1

(ii) Why would this polymer be formed as a fibre and not as a resin?

1

(4)

Marks

6. A student used the simple laboratory apparatus shown to determine the enthalpy of combustion of methanol.

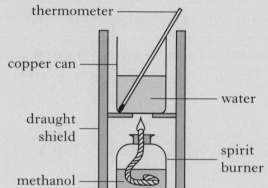

(a) (i) What measurements are needed to calculate the energy released by the burning methanol?

1

(ii) The student found that burning 0·370 g of methanol produces 3·86 kJ of energy.

Use this result to calculate the enthalpy of combustion of methanol.

1

(b) A more accurate value can be obtained using a bomb calorimeter.

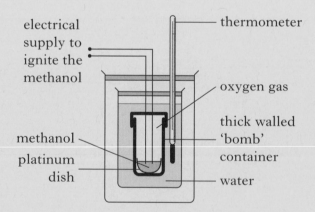

One reason for the more accurate value is that less heat is lost to the surroundings than in the simple laboratory method.

Give **one** other reason for the value being more accurate in the bomb calorimeter method.

1

Marks

7.   An experiment was carried out to determine the rate of the reaction between hydrochloric acid and calcium carbonate chips.   The rate of this reaction was followed by measuring the volume of gas released over a certain time.

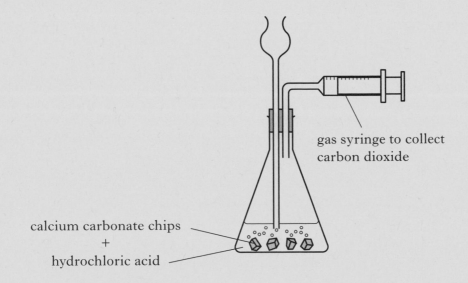

gas syringe to collect carbon dioxide

calcium carbonate chips
+
hydrochloric acid

(a) Describe a different way of measuring volume in order to follow the rate of this reaction.

1

(b) What other variable could be measured to follow the rate of this reaction?

1

(2)

[Turn over

Marks

8. Ammonia is produced in industry by the Haber Process.

$$N_2(g) + 3H_2(g) \rightleftharpoons 2NH_3(g)$$

(*a*) State whether the industrial manufacture of ammonia is likely to be a batch or a continuous process.

1

(*b*) The graph shows how the percentage yield of ammonia changes with temperature at a pressure of 100 atmospheres.

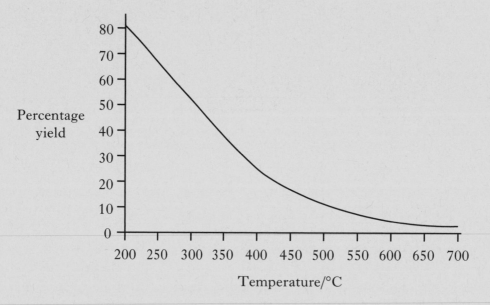

(i) A student correctly concludes from the graph that the production of ammonia is an exothermic process.

What is the reasoning that leads to this conclusion?

1

(ii) **Explain clearly** why the industrial manufacture of ammonia is carried out at a pressure greater than 100 atmospheres.

2

Marks

8. **(continued)**

   (*c*) Under certain conditions, 500 kg of nitrogen reacts with excess hydrogen to produce 405 kg of ammonia.

   Calculate the percentage yield of ammonia under these conditions.

   Show your working clearly.

2

(6)

[Turn over

DO NOT WRITE IN THIS MARGIN

Marks

9. Primary, secondary and tertiary alkanols can be prepared by the reaction of carbonyl compounds with Grignard reagents.

Step 1

The Grignard reagent reacts with the carbonyl compound.

$$CH_3-CH_2-CH_2-C{\overset{O}{\underset{H}{}}} \quad + \quad CH_3MgCl \quad \longrightarrow \quad CH_3-CH_2-CH_2-\underset{\underset{CH_3}{|}}{\overset{\overset{OMgCl}{|}}{C}}-H$$

butanal   Grignard reagent

Step 2

The reaction of the product of **Step 1** with dilute acid produces the alkanol.

$$CH_3-CH_2-CH_2-\underset{\underset{CH_3}{|}}{\overset{\overset{OMgCl}{|}}{C}}-H \quad \longrightarrow \quad CH_3-CH_2-CH_2-\underset{\underset{CH_3}{|}}{\overset{\overset{OH}{|}}{C}}-H$$

    + HCl   + MgCl$_2$

(a) Describe the difference between a primary, a secondary and a tertiary alkanol. You may wish to include labelled structures in your answer.

1

(b) Suggest a name for the type of reaction that takes place in **Step 1**.

1

Marks

9. (continued)

(c)   The same Grignard reagent can be used to produce the alkanol below.

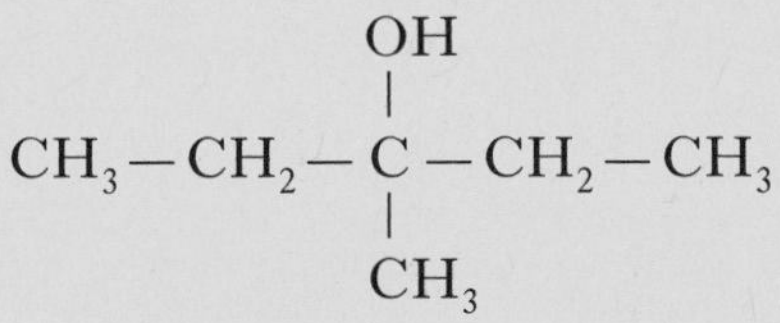

$$CH_3 - CH_2 - \overset{\displaystyle OH}{\underset{\displaystyle CH_3}{\overset{|}{\underset{|}{C}}}} - CH_2 - CH_3$$

Name the carbonyl compound used in this reaction.

1
(3)

[Turn over

DO NOT WRITE IN THIS MARGIN

Marks

10. Sherbet contains a mixture of sodium hydrogencarbonate and tartaric acid. The fizzing sensation in the mouth is due to the carbon dioxide produced in the following reaction.

$2NaHCO_3$	+	$C_4H_6O_6$	$\rightarrow$	$Na_2(C_4H_4O_6)$	+	$2H_2O$	+	$2CO_2$
sodium hydrogencarbonate		tartaric acid		sodium tartrate				

(*a*) Name the type of reaction taking place. **1**

(*b*) The chemical name for tartaric acid is 2,3-dihydroxybutanedioic acid.

Draw a structural formula for tartaric acid. **1**

(*c*) In an experiment, a student found that adding water to 20 sherbet sweets produced 105 cm^3 of carbon dioxide.

Assuming that sodium hydrogencarbonate is in excess, calculate the average mass of tartaric acid, in grams, in one sweet.

(Take the molar volume of carbon dioxide to be 24 litre mol^{-1}.)

Show your working clearly. **2**

(4)

Marks

11. The following answers were taken from a student's examination paper.

The two answers are incorrect.

For each question, give the correct explanation.

(*a*) **Question**   As a rough guide, the rate of a reaction tends to double for every $10\,^{\circ}\mathrm{C}$ rise in temperature.

Why does a small increase in temperature produce a large increase in reaction rate?

Student answer   Because rising temperature increases the activation energy which increases the number of collisions which speeds up the reaction greatly.

Correct explanation

1

(*b*) **Question**   Explain the difference in atomic size between potassium and chlorine atoms.

Student answer   A potassium nucleus has 19 protons but a chlorine nucleus has only 17 protons. The greater pull on the outer electron in the potassium atom means the atomic size of potassium is less than that of chlorine.

Correct explanation

1

(2)

[Turn over

DO NOT
WRITE IN
THIS
MARGIN

12. Barium hydroxide solution neutralises dilute sulphuric acid. A white precipitate of barium sulphate is formed in the reaction.

$$H_2SO_4(aq) \quad + \quad Ba(OH)_2(aq) \quad \rightarrow \quad BaSO_4(s) \quad + \quad 2H_2O(\ell)$$

The reaction can be followed by measuring the conductivity of the solution as barium hydroxide solution is added to dilute sulphuric acid.

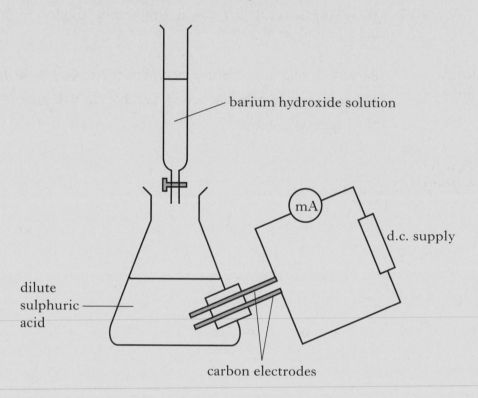

The graph shows how the conductivity changed in one experiment when barium hydroxide solution was added to $50\,cm^3$ of $0 \cdot 01\ mol\,l^{-1}$ sulphuric acid.

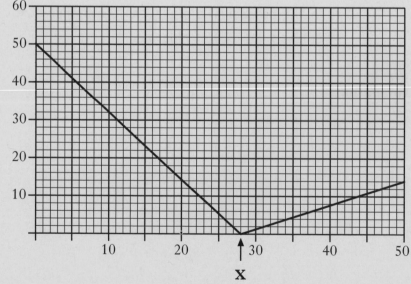

DO NOT WRITE IN THIS MARGIN

Marks

12. (continued)

(*a*) Point **X** corresponds to the end-point of the reaction.

Use the information on the graph to calculate the concentration of the barium hydroxide solution, in $mol\,l^{-1}$. **1**

(*b*) **Explain clearly** why the conductivity is very close to zero at the end-point. **2**

(3)

[Turn over

Marks

13. An electrolysis experiment was set up with two cells as shown.

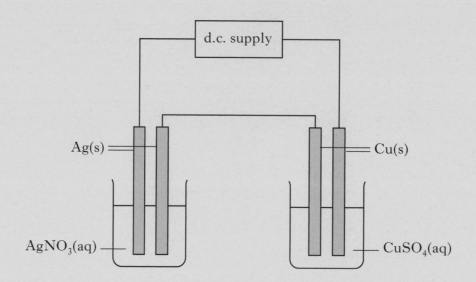

The reaction at the silver electrode is:

$$Ag^+(aq) \quad + \quad e^- \quad \rightarrow \quad Ag(s)$$

The mass of silver deposited in the reaction was $0.365\,g$.

(a) In addition to measuring the time, what **two** changes to the circuit would need to be made to find accurately the quantity of charge required to deposit $0.365\,g$ of silver?

1

(b) What mass of copper would be deposited in the same time?

Show your working clearly.

2

(3)

Marks

14. (a) A student compared the properties of equal concentrations of aqueous solutions of hydrochloric acid and ethanoic acid.

Experiment	Hydrochloric acid	Ethanoic acid
1 Rate of reaction with magnesium	fast	faster (slower) same
2 Electrical conductivity	80 mA	higher lower same
3 Volume of 0·1 mol l^{-1} sodium hydroxide to neutralise 20 cm^3 acid	20 cm^3	more less same

The result for ethanoic acid has been circled for experiment 1.

Circle the expected results for ethanoic acid in experiments 2 and 3.

1

(b) Some of the hydrogen atoms in ethanoic acid can be replaced by chlorine atoms to give three chloroethanoic acids. The student measured the pH of aqueous solutions of the four related acids. All the acids had the same concentration.

Acid name	Molecular formula	pH
ethanoic acid	CH_3COOH	2·3
chloroethanoic acid	$CH_2ClCOOH$	1·4
dichloroethanoic acid	$CHCl_2COOH$	1·0
trichloroethanoic acid	CCl_3COOH	0·2

(i) What is the concentration of hydroxide ions, in mol l^{-1}, in the dichloroethanoic acid solution?

1

(ii) **Explain clearly** the way in which the number of chlorine atoms in the acid molecules affects the strength of the acid.

2

[Turn over   (4)

Marks

15. (*a*) Methane is produced in the reaction of aluminium carbide with water.

$$Al_4C_3 \quad + \quad H_2O \quad \rightarrow \quad Al(OH)_3 \quad + \quad CH_4$$

Balance the above equation.

1

(*b*) Silane, silicon hydride, is formed in the reaction of silicon with hydrogen.

$$Si(s) \quad + \quad 2H_2(g) \quad \rightarrow \quad SiH_4(g)$$
$$\text{silane}$$

The enthalpy change for this reaction is called the enthalpy of formation of silane.

The combustion of silane gives silicon dioxide and water.

$$SiH_4(g) \quad + \quad 2O_2(g) \quad \rightarrow \quad SiO_2(s) \quad + \quad 2H_2O(\ell) \quad \Delta H = -1517\,kJ\,mol^{-1}$$

The enthalpy of combustion of silicon is $-911\,kJ\,mol^{-1}$.

Use this information and the enthalpy of combustion of hydrogen in the data booklet to calculate the enthalpy of formation of silane, in $kJ\,mol^{-1}$.

Show your working clearly.

2

(3)

Marks

16. Thorium-227 decays by alpha emission.

(*a*) Complete the nuclear equation for the alpha decay of thorium-227.

$$^{227}Th \quad \rightarrow$$

1

(*b*) A sample of thorium-227 was placed in a wooden box. A radiation detector was held 10 cm away from the box.

Why was alpha radiation not detected?

1

(*c*) Thorium-227 has a half-life of 19 days. If 0·42 g of thorium-227 has decayed after 57 days, calculate the initial mass of thorium-227, in grams.

1

(3)

[Turn over

DO NOT WRITE IN THIS MARGIN

Marks

17. Carbon-13 NMR is a technique used in chemistry to determine the structure of organic compounds.

(*a*) Calculate the neutron to proton ratio in an atom of carbon-13. **1**

(*b*) The technique allows a carbon atom in a molecule to be identified by its 'chemical shift'. This value depends on the other atoms bonded to the carbon atom.

Shift table

Carbon environment	Chemical shift/ppm
C = O (in ketones)	205 – 220
C = O (in aldehydes)	190 – 205
C = O (in acids and esters)	170 – 185
C = **C** (in alkenes)	115 – 140
C ≡ **C** (in alkynes)	70 – 95
—**C**H	25 – 50
—$\mathbf{C}H_2$	16 – 40
—$\mathbf{C}H_3$	5 – 15

In a carbon-13 NMR spectrum, the number of lines correspond to the number of chemically different carbon atoms and the position of the line (the value of the chemical shift) indicates the type of carbon atom.

DO NOT
WRITE IN
THIS
MARGIN

Marks

17. (b) (continued)

The spectrum for propanal is shown.

Spectrum 1

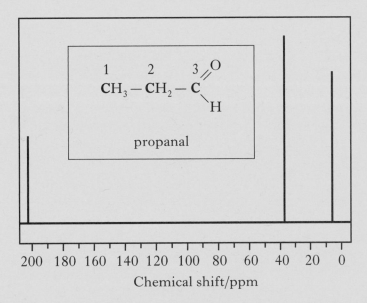

(i) Use the table of chemical shifts to label each of the peaks on the spectrum with a number to match the carbon atom in propanal that is responsible for the peak.

1

(ii) Hydrocarbon **X** has a relative formula mass of 54. Hydrocarbon **X** reacts with hydrogen. One of the products, hydrocarbon **Y**, has a relative formula mass of 56.

The carbon-13 NMR spectrum for hydrocarbon **Y** is shown below.

Spectrum 2

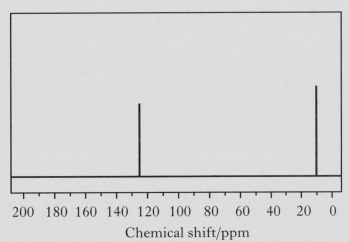

Name hydrocarbon **Y**.

1

(3)

Marks

18. The number of moles of carbon monoxide in a sample of air can be measured as follows.

   Step 1 The carbon monoxide reacts with iodine(V) oxide, producing iodine.

   $$5CO(g) + I_2O_5(s) \rightarrow I_2(s) + 5CO_2(g)$$

   Step 2 The iodine is then dissolved in potassium iodide solution and titrated against sodium thiosulphate solution.

   $$I_2(aq) + 2S_2O_3^{2-}(aq) \rightarrow S_4O_6^{2-}(aq) + 2I^-(aq)$$

   (a) Write the ion-electron equation for the oxidation reaction in **Step 2**.

   1

   (b) Name a chemical that can be used to indicate when all of the iodine has been removed in the reaction taking place in **Step 2**.

   1

   (c) If $50 \cdot 4 \, cm^3$ of $0 \cdot 10 \, mol \, l^{-1}$ sodium thiosulphate solution was used in a titration, calculate the number of moles of carbon monoxide in the sample of air.

   Show your working clearly.

   2
   (4)

[*END OF QUESTION PAPER*]

ADDITIONAL SPACE FOR ANSWERS

ADDITIONAL SPACE FOR ANSWERS

DO NOT
WRITE IN
THIS
MARGIN

ADDITIONAL SPACE FOR ANSWERS

ADDITIONAL SPACE FOR ANSWERS

[BLANK PAGE]

[BLANK PAGE]

FOR OFFICIAL USE

Total
Section B

X012/301

NATIONAL
QUALIFICATIONS
2010

WEDNESDAY, 2 JUNE
9.00 AM – 11.30 AM

CHEMISTRY
HIGHER

Fill in these boxes and read what is printed below.

Full name of centre

Town

Forename(s)

Surname

Date of birth

Day   Month   Year   Scottish candidate number   Number of seat

Reference may be made to the Chemistry Higher and Advanced Higher Data Booklet.

SECTION A   Questions 1   40 (40 marks)

Instructions for completion of **Section A** are given on page two.

For this section of the examination you must use an **HB pencil**.

SECTION B (60 marks)

1   All questions should be attempted.

2   The questions may be answered in any order but all answers are to be written in the spaces provided in this answer book, **and must be written clearly and legibly in ink**.

3   Rough work, if any should be necessary, should be written in this book and then scored through when the fair copy has been written. If further space is required, a supplementary sheet for rough work may be obtained from the Invigilator.

4   Additional space for answers will be found at the end of the book. If further space is required, supplementary sheets may be obtained from the Invigilator and should be inserted inside the **front** cover of this book.

5   The size of the space provided for an answer should not be taken as an indication of how much to write. It is not necessary to use all the space.

6   Before leaving the examination room you must give this book to the Invigilator. If you do not, you may lose all the marks for this paper.

SECTION A

Read carefully

1 Check that the answer sheet provided is for **Chemistry Higher (Section A)**.

2 For this section of the examination you must use an **HB pencil** and, where necessary, an eraser.

3 Check that the answer sheet you have been given has **your name**, **date of birth**, **SCN** (Scottish Candidate Number) and **Centre Name** printed on it.

Do not change any of these details.

4 If any of this information is wrong, tell the Invigilator immediately.

5 If this information is correct, **print** your name and seat number in the boxes provided.

6 The answer to each question is **either** A, B, C or D. Decide what your answer is, then, using your pencil, put a horizontal line in the space provided (see sample question below).

7 There is **only one correct** answer to each question.

8 Any rough working should be done on the question paper or the rough working sheet, **not** on your answer sheet.

9 At the end of the examination, put the **answer sheet for Section A inside the front cover of your answer book**.

Sample Question

To show that the ink in a ball-pen consists of a mixture of dyes, the method of separation would be

     A   chromatography

     B   fractional distillation

     C   fractional crystallisation

     D   filtration.

The correct answer is **A**—chromatography. The answer **A** has been clearly marked in **pencil** with a horizontal line (see below).

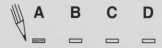

Changing an answer

If you decide to change your answer, carefully erase your first answer and using your pencil, fill in the answer you want. The answer below has been changed to **D**.

A   B   C   D

1. Which of the following gases would dissolve in water to form an alkali?

   A HBr

   B NH_3

   C CO_2

   D CH_4

2. Which of the following pairs of solutions is most likely to produce a precipitate when mixed?

   A Magnesium nitrate + sodium chloride

   B Magnesium nitrate + sodium sulphate

   C Silver nitrate + sodium chloride

   D Silver nitrate + sodium sulphate

3. 0·5 mol of copper(II) chloride and 0·5 mol of copper(II) sulphate are dissolved together in water and made up to 500 cm^3 of solution.

   What is the concentration of Cu^{2+}(aq) ions in the solution in $mol\,l^{-1}$?

   A 0·5

   B 1·0

   C 2·0

   D 4·0

4. For any chemical, its temperature is a measure of

   A the average kinetic energy of the particles that react

   B the average kinetic energy of all the particles

   C the activation energy

   D the minimum kinetic energy required before reaction occurs.

5. 1 mol of hydrogen gas and 1 mol of iodine vapour were mixed and allowed to react. After t seconds, 0·8 mol of hydrogen remained.

   The number of moles of hydrogen iodide formed at t seconds was

   A 0·2

   B 0·4

   C 0·8

   D 1·6.

6. Excess zinc was added to 100 cm^3 of hydrochloric acid, concentration 1 $mol\,l^{-1}$.

   Graph I refers to this reaction.

   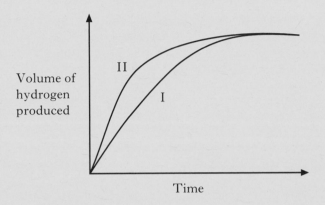

   Graph II could be for

   A excess zinc reacting with 100 cm^3 of hydrochloric acid, concentration 2 $mol\,l^{-1}$

   B excess zinc reacting with 100 cm^3 of sulphuric acid, concentration 1 $mol\,l^{-1}$

   C excess zinc reacting with 100 cm^3 of ethanoic acid, concentration 1 $mol\,l^{-1}$

   D excess magnesium reacting with 100 cm^3 of hydrochloric acid, concentration 1 $mol\,l^{-1}$.

7. Which of the following is **not** a correct statement about the effect of a catalyst?

   The catalyst

   A provides an alternative route to the products

   B lowers the energy that molecules need for successful collisions

   C provides energy so that more molecules have successful collisions

   D forms bonds with reacting molecules.

8. A potential energy diagram can be used to show the activation energy (E_A) and the enthalpy change (ΔH) for a reaction.

   Which of the following combinations of E_A and ΔH could **never** be obtained for a reaction?

   A $E_A = 50\,kJ\,mol^{-1}$ and $\Delta H = -100\,kJ\,mol^{-1}$

   B $E_A = 50\,kJ\,mol^{-1}$ and $\Delta H = +100\,kJ\,mol^{-1}$

   C $E_A = 100\,kJ\,mol^{-1}$ and $\Delta H = +50\,kJ\,mol^{-1}$

   D $E_A = 100\,kJ\,mol^{-1}$ and $\Delta H = -50\,kJ\,mol^{-1}$

[Turn over

9. As the relative atomic mass in the halogens increases

A the boiling point increases

B the density decreases

C the first ionisation energy increases

D the atomic size decreases.

10. The table shows the first three ionisation energies of aluminium.

Ionisation energy/kJ mol^{-1}		
1st	2nd	3rd
584	1830	2760

Using this information, what is the enthalpy change, in kJ mol^{-1}, for the following reaction?

$$Al^{3+}(g) + 2e^- \rightarrow Al^+(g)$$

A +2176

B −2176

C +4590

D −4590

11. When two atoms form a non-polar covalent bond, the two atoms **must** have

A the same atomic size

B the same electronegativity

C the same ionisation energy

D the same number of outer electrons.

12. In which of the following liquids does hydrogen bonding occur?

A Ethanoic acid

B Ethyl ethanoate

C Hexane

D Hex-1-ene

13. Which line in the table shows the correct entries for tetrafluoroethene?

	Polar bonds?	Polar molecules?
A	yes	yes
B	yes	no
C	no	no
D	no	yes

14. Element **X** was found to have the following properties.

(i) It does not conduct electricity when solid.

(ii) It forms a gaseous oxide.

(iii) It is a solid at room temperature.

Element **X** could be

A magnesium

B silicon

C nitrogen

D sulphur.

15. The Avogadro Constant is the same as the number of

A molecules in 16 g of oxygen

B ions in 1 litre of sodium chloride solution, concentration 1 mol l^{-1}

C atoms in 24 g of carbon

D molecules in 2 g of hydrogen.

16. Which of the following contains one mole of neutrons?

A 1 g of $^{1}_{1}H$

B 1 g of $^{12}_{6}C$

C 2 g of $^{24}_{12}Mg$

D 2 g of $^{22}_{10}Ne$

17. $20 \, cm^3$ of ammonia gas reacted with an excess of heated copper(II) oxide.

$$3CuO + 2NH_3 \rightarrow 3Cu + 3H_2O + N_2$$

Assuming all measurements were made at $200\,°C$, what would be the volume of gaseous products?

A $10 \, cm^3$

B $20 \, cm^3$

C $30 \, cm^3$

D $40 \, cm^3$

18. Which of the following fuels can be produced by the fermentation of biological material under anaerobic conditions?

A Methane

B Ethane

C Propane

D Butane

19. Rum flavouring is based on the compound with the formula shown.

$$CH_3CH_2CH_2C \underset{OCH_2CH_3}{\overset{O}{<}}$$

It can be made from

A ethanol and butanoic acid

B propanol and ethanoic acid

C butanol and methanoic acid

D propanol and propanoic acid.

20. Which of the following structural formulae represents a tertiary alcohol?

A
$$CH_3 - \overset{\overset{\displaystyle CH_3}{|}}{\underset{\underset{\displaystyle CH_3}{|}}{C}} - CH_2 - OH$$

B
$$CH_3 - \overset{\overset{\displaystyle CH_3}{|}}{\underset{\underset{\displaystyle OH}{|}}{C}} - CH_2 - CH_3$$

C
$$CH_3 - CH_2 - CH_2 - \overset{\overset{\displaystyle H}{|}}{\underset{\underset{\displaystyle OH}{|}}{C}} - CH_3$$

D
$$CH_3 - CH_2 - \overset{\overset{\displaystyle H}{|}}{\underset{\underset{\displaystyle OH}{|}}{C}} - CH_2 - CH_3$$

21. What is the product when one mole of chlorine gas reacts with one mole of ethyne?

A 1,1-Dichloroethene

B 1,1-Dichloroethane

C 1,2-Dichloroethene

D 1,2-Dichloroethane

[Turn over

22.

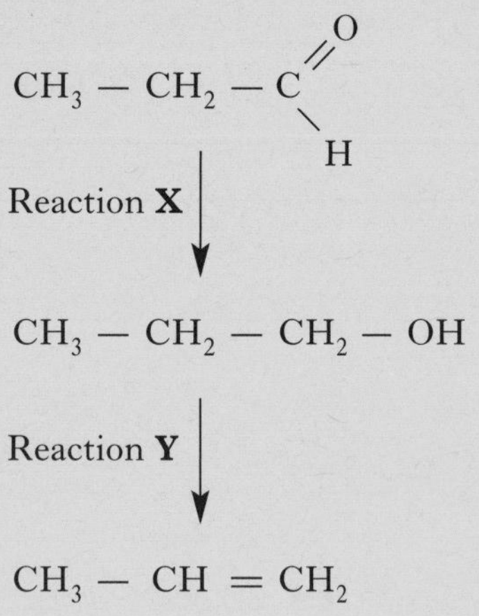

$$CH_3 - CH_2 - C \overset{O}{\underset{H}{\diagdown}}$$

Reaction **X** ↓

$$CH_3 - CH_2 - CH_2 - OH$$

Reaction **Y** ↓

$$CH_3 - CH = CH_2$$

Which line in the table correctly describes reactions **X** and **Y**?

	Reaction X	Reaction Y
A	oxidation	dehydration
B	oxidation	condensation
C	reduction	dehydration
D	reduction	condensation

23. Ozone has an important role in the upper atmosphere because it

A   absorbs ultraviolet radiation

B   absorbs certain CFCs

C   reflects ultraviolet radiation

D   reflects certain CFCs.

24. Synthesis gas consists mainly of

A CH_4 alone

B CH_4 and CO

C   CO and H_2

D CH_4, CO and H_2.

25. Ethene is used in the manufacture of addition polymers.

What type of reaction is used to produce ethene from ethane?

A   Cracking

B   Addition

C   Oxidation

D   Hydrogenation

26. Polyester fibres and cured polyester resins are both very strong.

Which line in the table correctly describes the structure of these polyesters?

	Fibre	Cured resin
A	cross-linked	cross-linked
B	linear	linear
C	cross-linked	linear
D	linear	cross-linked

27. Part of a polymer chain is shown below.

$$- O - \overset{O}{\overset{\|}{C}} - (CH_2)_4 - \overset{O}{\overset{\|}{C}} - O - (CH_2)_6 - O - \overset{O}{\overset{\|}{C}} - (CH_2)_4 - \overset{O}{\overset{\|}{C}} - O - (CH_2)_6 - O -$$

Which of the following compounds, when added to the reactants during polymerisation, would stop the polymer chain from getting too long?

A $HO - \overset{O}{\overset{\|}{C}} - (CH_2)_4 - \overset{O}{\overset{\|}{C}} - OH$

B $HO - (CH_2)_6 - OH$

C $HO - (CH_2)_5 - \overset{O}{\overset{\|}{C}} - OH$

D $CH_3 - (CH_2)_4 - CH_2 - OH$

28. Which of the following fatty acids is unsaturated?

   A $C_{19}H_{39}COOH$

   B $C_{21}H_{43}COOH$

   C $C_{17}H_{31}COOH$

   D $C_{13}H_{27}COOH$

29. Which of the following alcohols is likely to be obtained on hydrolysis of butter?

   A $CH_3 - CH_2 - CH_2 - OH$

   B $CH_3 - CH - CH_3$
             |
             OH

   C $CH_2 - OH$
      |
      CH_2
      |
      $CH_2 - OH$

   D $CH_2 - OH$
      |
      $CH - OH$
      |
      $CH_2 - OH$

30. Amino acids are converted into proteins by

   A   hydration

   B   hydrolysis

   C   hydrogenation

   D   condensation.

31. Which of the following compounds is a raw material in the chemical industry?

   A   Ammonia

   B   Calcium carbonate

   C   Hexane

   D   Nitric acid

32. Given the equations

   $$Mg(s) + 2H^+(aq) \rightarrow Mg^{2+}(aq) + H_2(g)$$
   $$\Delta H = a \, J \, mol^{-1}$$

   $$Zn(s) + 2H^+(aq) \rightarrow Zn^{2+}(aq) + H_2(g)$$
   $$\Delta H = b \, J \, mol^{-1}$$

   $$Mg(s) + Zn^{2+}(aq) \rightarrow Mg^{2+}(aq) + Zn(s)$$
   $$\Delta H = c \, J \, mol^{-1}$$

   then, according to Hess's Law

   A $c = a - b$

   B $c = a + b$

   C $c = b - a$

   D $c = -b - a.$

33. In which of the following reactions would an increase in pressure cause the equilibrium position to move to the left?

   A $CO(g) + H_2O(g) \rightleftharpoons CO_2(g) + H_2(g)$

   B $CH_4(g) + H_2O(g) \rightleftharpoons CO(g) + 3H_2(g)$

   C $Fe_2O_3(s) + 3CO(g) \rightleftharpoons 2Fe(s) + 3CO_2(g)$

   D $N_2(g) + 3H_2(g) \rightleftharpoons 2NH_3(g)$

34. If ammonia is added to a solution containing copper(II) ions an equilibrium is set up.

   $$Cu^{2+}(aq) + 2OH^-(aq) + 4NH_3(aq) \rightleftharpoons Cu(NH_3)_4(OH)_2(aq)$$
   (deep blue)

   If acid is added to this equilibrium system

   A   the intensity of the deep blue colour will increase

   B   the equilibrium position will move to the right

   C   the concentration of $Cu^{2+}(aq)$ ions will increase

   D   the equilibrium position will not be affected.

35. Which of the following is the best description of a $0 \cdot 1 \, mol \, l^{-1}$ solution of hydrochloric acid?

   A   Dilute solution of a weak acid

   B   Dilute solution of a strong acid

   C   Concentrated solution of a weak acid

   D   Concentrated solution of a strong acid

[Turn over

36. A solution has a negative pH value.

This solution

A neutralises $H^+(aq)$ ions

B contains no $OH^-(aq)$ ions

C has a high concentration of $H^+(aq)$ ions

D contains neither $H^+(aq)$ ions nor $OH^-(aq)$ ions.

37. When a certain aqueous solution is diluted, its conductivity decreases but its pH remains constant.

It could be

A ethanoic acid

B sodium chloride

C sodium hydroxide

D nitric acid.

38. Equal volumes of four $1 \, mol\,l^{-1}$ solutions were compared.

Which of the following $1 \, mol\,l^{-1}$ solutions contains the most ions?

A Nitric acid

B Hydrochloric acid

C Ethanoic acid

D Sulphuric acid

39. In which reaction is hydrogen gas acting as an oxidising agent?

A $H_2 + CuO \rightarrow H_2O + Cu$

B $H_2 + C_2H_4 \rightarrow C_2H_6$

C $H_2 + Cl_2 \rightarrow 2HCl$

D $H_2 + 2Na \rightarrow 2NaH$

40. Which particle will be formed when an atom of $^{211}_{83}Bi$ emits an α-particle and the decay product then emits a β-particle?

A $^{207}_{82}Pb$

B $^{208}_{81}Tl$

C $^{209}_{80}Hg$

D $^{210}_{79}Au$

Candidates are reminded that the answer sheet MUST be returned INSIDE the front cover of this answer book.

DO NOT
WRITE IN
THIS
MARGIN

Marks

SECTION B

All answers must be written clearly and legibly in ink.

1. The elements lithium, boron and nitrogen are in the second period of the Periodic Table.

   Complete the table below to show **both** the bonding and structure of these three elements at room temperature.

Name of element	Bonding	Structure
lithium		lattice
boron		
nitrogen	covalent	

2
(2)

[Turn over

Marks

2. (*a*) Polyhydroxyamide is a recently developed fire-resistant polymer.

The monomers used to produce the polymer are shown.

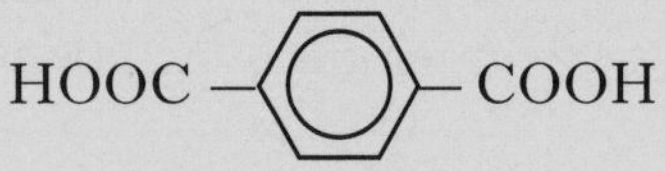

HOOC —⟨O⟩— COOH   H₂N —⟨O⟩— NH₂
                                        |
                                        OH

diacid 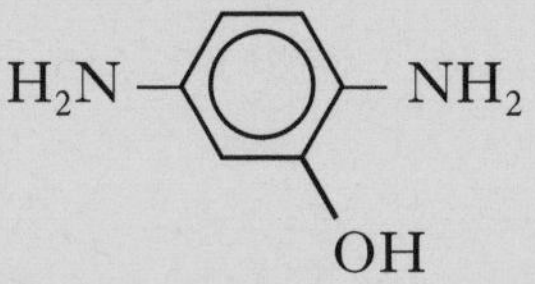   diamine

(i) How many hydrogen atoms are present in a molecule of the diamine molecule?

1

(ii) Draw a section of polyhydroxyamide showing **one** molecule of each monomer joined together.

1

(*b*) Poly(ethenol), another recently developed polymer, has an unusual property for a plastic.

What is this unusual property?

1
(3)

Marks

3. Atmospheric oxygen, $O_2(g)$, dissolves in the Earth's oceans forming dissolved oxygen, $O_2(aq)$, which is essential for aquatic life.

An equilibrium is established.

$$O_2(g) \quad + \quad (aq) \quad \rightleftharpoons \quad O_2(aq) \qquad \Delta H = -12 \cdot 1 \, kJ \, mol^{-1}$$

(a) (i) What is meant by a reaction at "equilibrium"?

1

(ii) What would happen to the concentration of dissolved oxygen if the temperature of the Earth's oceans increased?

1

(b) A sample of oceanic water was found to contain 0·010 g of dissolved oxygen.

Calculate the number of moles of dissolved oxygen present in the sample.

1

(3)

Marks

4. In the Hall-Heroult Process, aluminium is produced by the electrolysis of an ore containing aluminium oxide.

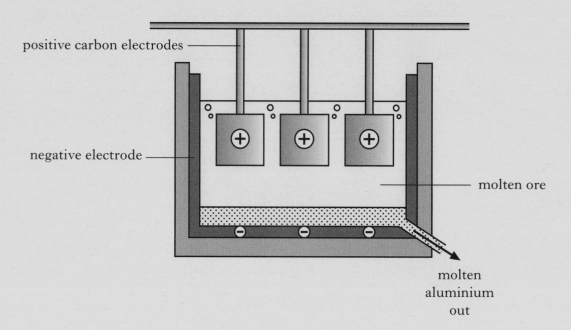

positive carbon electrodes

negative electrode

molten ore

molten
aluminium
out

(*a*) Suggest why the positive carbon electrodes need to be replaced regularly.

1

(*b*) Calculate the mass of aluminium, in grams, produced in 20 minutes when a current of 50 000 A is used.

Show your working clearly.

3

(4)

Marks

5. The reaction of oxalic acid with an acidified solution of potassium permanganate was studied to determine the effect of temperature changes on reaction rate.

$$5(COOH)_2(aq) + 6H^+(aq) + 2MnO_4^-(aq) \rightarrow 2Mn^{2+}(aq) + 10CO_2(g) + 8H_2O(\ell)$$

The reaction was carried out at several temperatures between 40 °C and 60 °C. The end of the reaction was indicated by a colour change from purple to colourless.

(a) (i) State **two** factors that should be kept the same in these experiments.

1

(ii) Why is it difficult to measure an accurate value for the reaction time when the reaction is carried out at room temperature?

1

(b) Sketch a graph to show how the rate varied with increasing temperature.

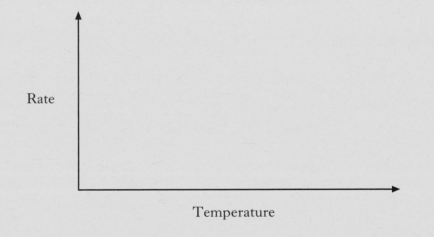

1

(3)

[Turn over

Marks

6. Positron emission tomography, PET, is a technique that provides information about biochemical processes in the body.

   Carbon-11, ^{11}C, is a positron-emitting radioisotope that is injected into the bloodstream.

   A positron can be represented as $^{0}_{1}e$.

   (a) Complete the nuclear equation for the decay of ^{11}C by positron-emission.

   $$^{11}C \longrightarrow$$

   1

   (b) A sample of ^{11}C had an initial count rate of 640 counts min^{-1}. After 1 hour the count rate had fallen to 80 counts min^{-1}.

   Calculate the half-life, in minutes, of ^{11}C.

   1

   (c) ^{11}C is injected into the bloodstream as glucose molecules ($C_6H_{12}O_6$). Some of the carbon atoms in these glucose molecules are ^{11}C atoms.

   The intensity of radiation in a sample of ^{11}C is compared with the intensity of radiation in a sample of glucose containing ^{11}C atoms. Both samples have the same mass.

   Which sample has the higher intensity of radiation?

   Give a reason for your answer.

   1

   (3)

Marks

7. Hydrogen cyanide, HCN, is highly toxic.

(*a*) Information about hydrogen cyanide is given in the table.

Structure	$H-C\equiv N$
Molecular mass	27
Boiling point	26 °C

Although hydrogen cyanide has a similar molecular mass to nitrogen, it has a much higher boiling point. This is due to the permanent dipole–permanent dipole attractions in liquid hydrogen cyanide.

What is meant by permanent dipole–permanent dipole attractions?

Explain how they arise in liquid hydrogen cyanide.

2

(*b*) Hydrogen cyanide is of great importance in organic chemistry. It offers a route to increasing the chain length of a molecule.

If ethanal is reacted with hydrogen cyanide and the product hydrolysed with acid, lactic acid is formed.

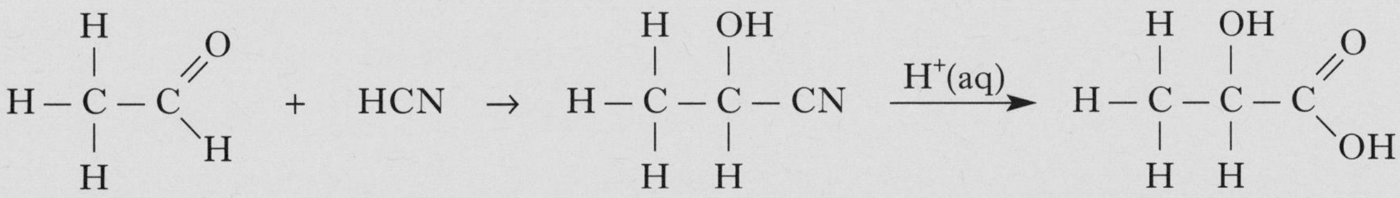

ethanal   lactic acid

Draw a structural formula for the acid produced when propanone is used instead of ethanal in the above reaction sequence.

1

(3)

Marks

8. Glycerol, $C_3H_8O_3$, is widely used as an ingredient in toothpaste and cosmetics.

(a) Glycerol is mainly manufactured from fats and oils. Propene can be used as a feedstock in an alternative process as shown.

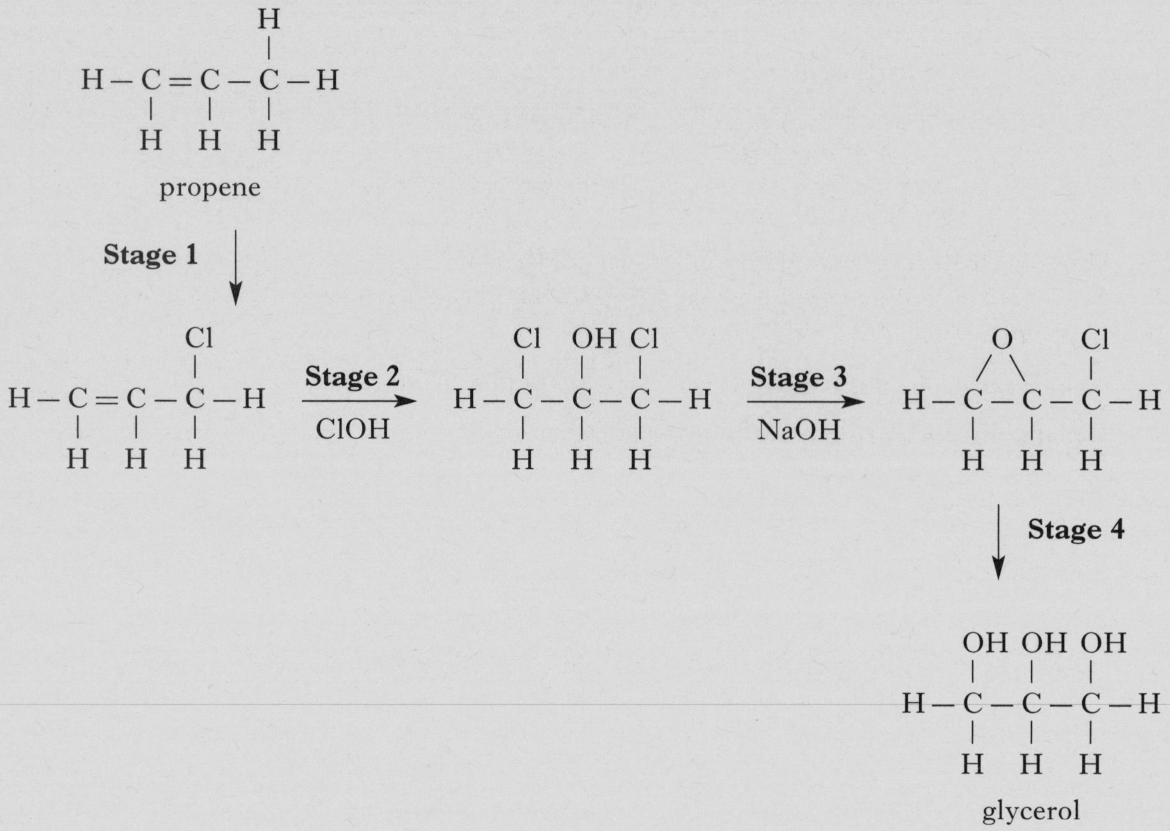

(i) What is meant by a feedstock?

1

(ii) Name the type of reaction taking place in **Stage 2**.

1

(iii) In **Stage 3**, a salt and water are produced as by-products.

Name the salt produced.

1

DO NOT WRITE IN THIS MARGIN

Marks

8. (*a*) (continued)

(iv) Apart from cost, state **one** advantage of using fats and oils rather than propene in the manufacture of glycerol. 1

(*b*) Hydrogen has been named as a 'fuel for the future'. In a recent article researchers reported success in making hydrogen from glycerol:

$$C_3H_8O_3(\ell) \rightarrow \quad CO_2(g) + \quad CH_4(g) + \quad H_2(g)$$

Balance this equation. 1

(*c*) The enthalpy of formation of glycerol is the enthalpy change for the reaction:

$$\underset{\text{(graphite)}}{3C(s)} + 4H_2(g) + 1\tfrac{1}{2}O_2(g) \rightarrow C_3H_8O_3(\ell)$$

Calculate the enthalpy of formation of glycerol, in kJ mol^{-1}, using information from the data booklet and the following data.

$$C_3H_8O_3(\ell) + 3\tfrac{1}{2}O_2(g) \rightarrow 3CO_2(g) + 4H_2O(\ell) \quad \Delta H = -1654\,\text{kJ mol}^{-1}$$

Show your working clearly. 2

(7)

[Turn over

DO NOT WRITE IN THIS MARGIN

Marks

9. Enzymes are biological catalysts.

(*a*) Name the **four** elements present in all enzymes. **1**

(*b*) The enzyme catalase, found in potatoes, can catalyse the decomposition of hydrogen peroxide.

$$2H_2O_2(aq) \rightarrow 2H_2O(\ell) + O_2(g)$$

A student carried out the Prescribed Practical Activity (PPA) to determine the effect of pH on enzyme activity.

Describe how the activity of the enzyme was measured in this PPA. **1**

(*c*) A student wrote the following **incorrect** statement.

When the temperature is increased, enzyme-catalysed reactions will always speed up because more molecules have kinetic energy greater than the activation energy.

Explain the mistake in the student's reasoning. **1**

(3)

Marks

10. Sulphur trioxide can be prepared in the laboratory by the reaction of sulphur dioxide with oxygen.

$$2SO_2(g) + O_2(g) \rightleftharpoons 2SO_3(g)$$

The sulphur dioxide and oxygen gases are dried by bubbling them through concentrated sulphuric acid. The reaction mixture is passed over heated vanadium(V) oxide.

Sulphur trioxide has a melting point of 17 °C. It is collected as a white crystalline solid.

(a) Complete the diagram to show how the reactant gases are dried and the product is collected.

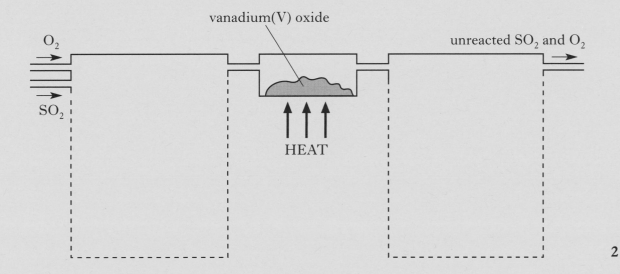

2

(b) Under certain conditions, 43·2 tonnes of sulphur trioxide are produced in the reaction of 51·2 tonnes of sulphur dioxide with excess oxygen.

Calculate the percentage yield of sulphur trioxide.

Show your working clearly.

2

(4)

[Turn over

Marks

11. (*a*) The first ionisation energy of an element is defined as the energy required to remove one mole of electrons from one mole of atoms in the gaseous state.

The graph shows the first ionisation energies of the Group 1 elements.

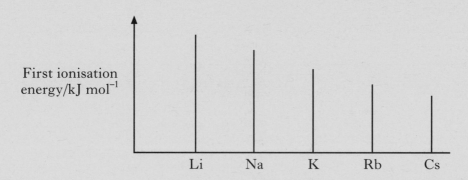

First ionisation
energy/kJ mol^{-1}

Li   Na   K   Rb   Cs

    (i) Clearly explain why the first ionisation energy decreases down this group.

2

    (ii) The energy needed to remove one electron from one helium atom is $3\cdot94 \times 10^{-21}$ kJ.

Calculate the first ionisation energy of helium, in kJ mol^{-1}.

1

(*b*) The ability of an atom to form a negative ion is measured by its Electron Affinity.

The Electron Affinity is defined as the energy change when one mole of gaseous atoms of an element combines with one mole of electrons to form gaseous negative ions.

Write the equation, showing state symbols, that represents the Electron Affinity of chlorine.

1

(4)

DO NOT
WRITE IN
THIS
MARGIN

Marks

12.   (*a*)   A student bubbled $240 \, cm^3$ of carbon dioxide into $400 \, cm^3$ of $0.10 \, mol \, l^{-1}$ lithium hydroxide solution.

The equation for the reaction is:

$$2LiOH(aq) \; + \; CO_2(g) \; \rightarrow \; Li_2CO_3(aq) \; + \; H_2O(\ell)$$

Calculate the number of moles of lithium hydroxide that would **not** have reacted.

(Take the molar volume of carbon dioxide to be 24 litres mol^{-1}.)

Show your working clearly.

2

(*b*)   What is the pH of the $0.10 \, mol \, l^{-1}$ lithium hydroxide solution used in the experiment?

1

(*c*)   Explain why lithium carbonate solution has a pH greater than 7.

In your answer you should mention the **two** equilibria involved.

2

(5)

Marks

13. (*a*) A sample of petrol was analysed to identify the hydrocarbons present. The results are shown in the table.

Number of carbon atoms per molecule	Hydrocarbons present in the sample
4	2-methylpropane
5	2-methylbutane
6	2,3-dimethylbutane
7	2,2-dimethylpentane 2,2,3-trimethylbutane

(i) Draw a structural formula for 2,2,3-trimethylbutane.

1

(ii) The structures of the hydrocarbons in the sample are similar in a number of ways.

What similarity in structure makes these hydrocarbons suitable for use in unleaded petrol?

1

(*b*) In some countries, organic compounds called 'oxygenates' are added to unleaded petrol.

One such compound is MTBE.

$$\text{MTBE} \qquad H_3C - \underset{\underset{CH_3}{|}}{\overset{\overset{CH_3}{|}}{C}} - O - CH_3$$

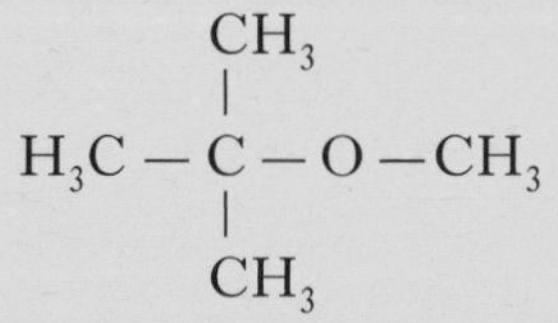

(i) Suggest why oxygenates such as MTBE are added to unleaded petrol.

1

Marks

13. **(b)** **(continued)**

(ii) MTBE is an example of an ether. All ethers contain the functional group:

$$-\overset{|}{\underset{|}{C}}-O-\overset{|}{\underset{|}{C}}-$$

Draw a structural formula for an isomer of MTBE that is also an ether.

1

(c) Some of the hydrocarbons that are suitable for unleaded petrol are produced by a process known as reforming.

One reforming reaction is:

$$H-\overset{\overset{\displaystyle H}{|}}{\underset{\underset{\displaystyle H}{|}}{C}}-\overset{\overset{\displaystyle H}{|}}{\underset{\underset{\displaystyle H}{|}}{C}}-\overset{\overset{\displaystyle H}{|}}{\underset{\underset{\displaystyle H}{|}}{C}}-\overset{\overset{\displaystyle H}{|}}{\underset{\underset{\displaystyle H}{|}}{C}}-\overset{\overset{\displaystyle H}{|}}{\underset{\underset{\displaystyle H}{|}}{C}}-\overset{\overset{\displaystyle H}{|}}{\underset{\underset{\displaystyle H}{|}}{C}}-H \quad \rightarrow \text{ hydrocarbon } \mathbf{A} \; + \; H_2$$

hexane

Hydrocarbon **A** is non-aromatic and does **not** decolourise bromine solution.

Give a possible name for hydrocarbon **A**.

1

(5)

[Turn over

Marks

14. (*a*) Hess's Law can be verified using the reactions summarised below.

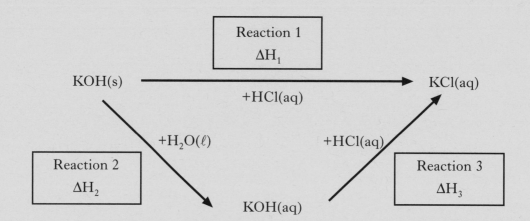

(i) Complete the list of measurements that would have to be carried out in order to determine the enthalpy change for Reaction 2.

> **Reaction 2**
>
> 1. Using a measuring cylinder, measure out $25\,cm^3$ of water into a polystyrene cup.
>
> 2.
>
> 3. Weigh out accurately about $1\cdot2\,g$ of potassium hydroxide and add it to the water, with stirring, until all the solid dissolves.
>
> 4.

1

(ii) Why was the reaction carried out in a polystyrene cup?

1

DO NOT WRITE IN THIS MARGIN

Marks

14. (*a*) (continued)

(iii) A student found that 1·08 kJ of energy was **released** when 1·2 g of potassium hydroxide was dissolved completely in water.

Calculate the enthalpy of solution of potassium hydroxide. **1**

(*b*) A student wrote the following **incorrect** statement.

The enthalpy of neutralisation for hydrochloric acid reacting with potassium hydroxide is less than that for sulphuric acid reacting with potassium hydroxide because fewer moles of water are formed as shown in these equations.

$$HCl + KOH \rightarrow KCl + H_2O$$

$$H_2SO_4 + 2KOH \rightarrow K_2SO_4 + 2H_2O$$

Explain the mistake in the student's statement. **1**

(4)

[Turn over

Marks

15. Infra-red spectroscopy is a technique that can be used to identify the bonds that are present in a molecule.

Different bonds absorb infra-red radiation of different wavenumbers. This is due to differences in the bond 'stretch'. These absorptions are recorded in a spectrum.

A spectrum for propan-1-ol is shown.

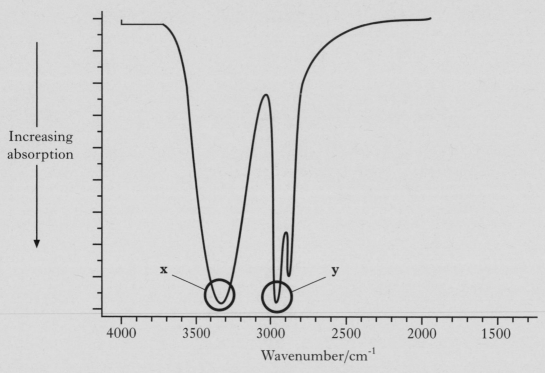

The correlation table on page 13 of the data booklet shows the wavenumber ranges for the absorptions due to different bonds.

(a) Use the correlation table to identify the bonds responsible for the two absorptions, **x** and **y**, that are circled in the propan-1-ol spectrum.

    x: **y:**

1

(b) Propan-1-ol reacts with ethanoic acid.

    (i) What name is given to this type of reaction?

1

Marks

15. (*b*) **(continued)**

(ii) Draw a spectrum that could be obtained for the organic product of this reaction.

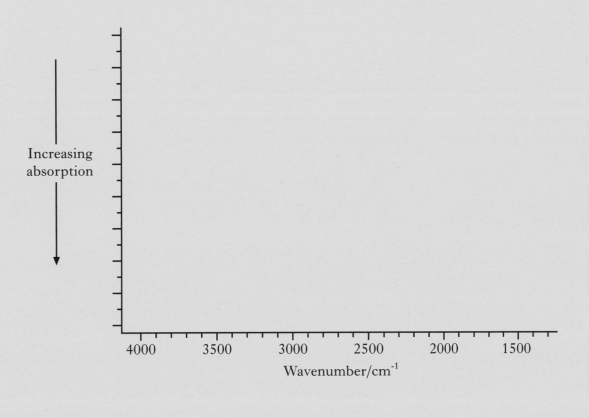

1

(3)

[Turn over

Marks

16. A major problem for the developed world is the pollution of rivers and streams by nitrite and nitrate ions.

The concentration of nitrite ions, $NO_2^-(aq)$, in water can be determined by titrating samples against acidified permanganate solution.

(*a*) Suggest **two** points of good practice that should be followed to ensure that an accurate end-point is achieved in a titration.

1

(*b*) An average of $21 \cdot 6 \, cm^3$ of $0 \cdot 0150 \, mol \, l^{-1}$ acidified permanganate solution was required to react completely with the nitrite ions in a $25 \cdot 0 \, cm^3$ sample of river water.

The equation for the reaction taking place is:

$$2MnO_4^-(aq) \ + \ 5NO_2^-(aq) \ + \ 6H^+(aq) \ \rightarrow \ 2Mn^{2+}(aq) \ + \ 5NO_3^-(aq) \ + \ 3H_2O(\ell)$$

(i) Calculate the nitrite ion concentration, in $mol \, l^{-1}$, in the river water.

Show your working clearly.

2

(ii) During the reaction the nitrite ion is oxidised to the nitrate ion.

Complete the ion-electron equation for the oxidation of the nitrite ions.

$$NO_2^-(aq) \quad \rightarrow \quad NO_3^-(aq)$$

1

[END OF QUESTION PAPER]

(4)

ADDITIONAL SPACE FOR ANSWERS

ADDITIONAL SPACE FOR ANSWERS

ADDITIONAL SPACE FOR ANSWERS

[BLANK PAGE]

[BLANK PAGE]

[BLANK PAGE]